It's another great book from CGP...

Physics exams can seem daunting — especially if you're not sure what to expect.
But they're less stressful if you've done plenty of realistic practice in advance.

Happily, this book (which includes a **free** Online Edition)
is packed with exam-style questions for every topic. It even includes
two complete practice exams to fully prepare you for the real thing.

How to get your free online edition

Want to read this book on your computer or tablet?
Just go to **cgpbooks.co.uk/extras** and enter this code...

1966 5563 0576 3753

By the way, this code only works for one person. If somebody else has used
this book before you, they might have already claimed the online edition.

CGP — still the best! ☺

Our sole aim here at CGP is to produce the highest quality books —
carefully written, immaculately presented and dangerously close to being funny.

Then we work our socks off to get them out to you
— at the cheapest possible prices.

Contents

☑ Use the tick boxes to check off the topics you've completed.

Section Three — Heating Processes

Section Four — Electricity

Section Five — Motors, Generators and Transformers

Section Six — Nuclear Physics

Practice Papers

How to get answers for the Practice Papers: Your free Online Edition of this book includes all the answers for Practice Papers 1 & 2. (Just flick back to the first page to find out how to get hold of your Online Edition.)

Published by CGP

Editors:
Jane Ellingham, Rachael Marshall, Matteo Orsini Jones, Charlotte Whiteley, Sarah Williams

With thanks to Ian Francis and Glenn Rogers for the proofreading.
With thanks to Mark Edwards for the reviewing.
With thanks to Laura Jakubowski for the copyright research.

ISBN: 978 1 84762 451 2

Data used to construct stopping distance graph on page 109 from the Highway Code.
© Crown Copyright re-produced under the terms of the Open Government licence
http://www.nationalarchives.gov.uk/doc/open-government-licence/

www.cgpbooks.co.uk

Clipart from Corel®
Printed by Elanders Ltd, Newcastle upon Tyne

Based on the classic CGP style created by Richard Parsons.

How to Use This Book

- Hold the book <u>upright</u>, approximately <u>50 cm</u> from your face, ensuring that the text looks like <u>this</u>, not s̄ı̄ɥ̄ʇ. Alternatively, place the book on a <u>horizontal</u> surface (e.g. a table or desk) and sit adjacent to the book, at a distance which doesn't make the text too small to read.

- In case of emergency, press the two halves of the book together <u>firmly</u> in order to close.

- Before attempting to use this book, familiarise yourself with the following <u>safety information</u>:

Just like in the real exams, some questions have a bit of blurb in italics at the start. This tells you that you'll be assessed on how well you write — so make sure your spelling, punctuation and grammar are all top notch.

The questions are arranged into topics, so you can get exam practice on exactly the bit of your course that you want.

Some questions have a bit of guidance or some working done for you, to help get you started on trickier topics. You won't get this in the exam though I'm afraid.

These contain handy tips to help you with specific questions.

There are answer lines for you to write your answers on. For calculation questions, there's also space for you to do your working.

You're told how many marks each question part is worth, and then the total for the whole question.

Exam Practice Tips give you hints to help with answering exam questions.

Use the answers at the back of the book to mark each page. Then you can find your score out of the total for the topic.

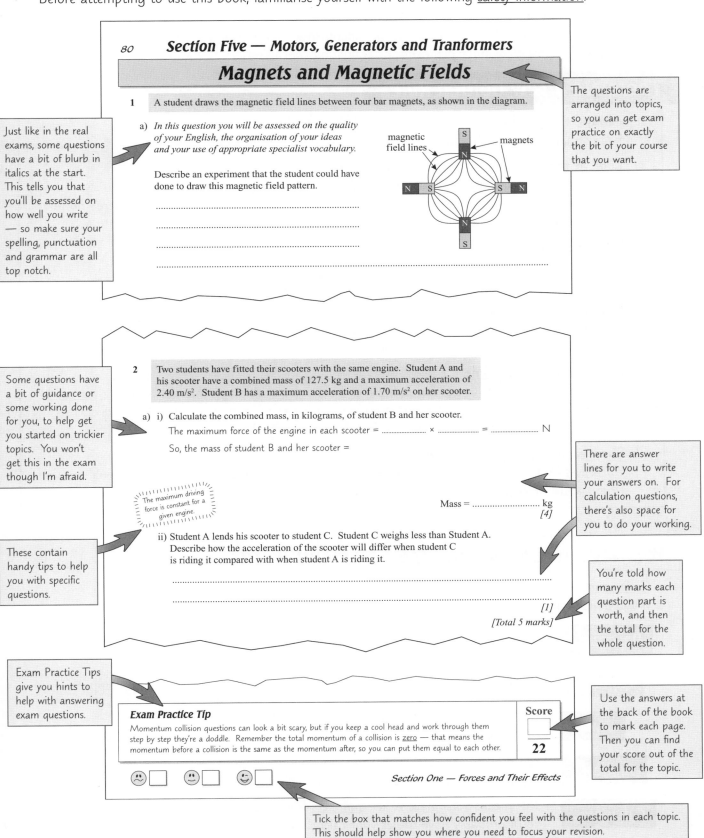

80   **Section Five — Motors, Generators and Tranformers**

Magnets and Magnetic Fields

1   A student draws the magnetic field lines between four bar magnets, as shown in the diagram.

a) *In this question you will be assessed on the quality of your English, the organisation of your ideas and your use of appropriate specialist vocabulary.*

Describe an experiment that the student could have done to draw this magnetic field pattern.

....

....

....

....

magnetic field lines   magnets

2   Two students have fitted their scooters with the same engine. Student A and his scooter have a combined mass of 127.5 kg and a maximum acceleration of 2.40 m/s². Student B has a maximum acceleration of 1.70 m/s² on her scooter.

a) i) Calculate the combined mass, in kilograms, of student B and her scooter.

The maximum force of the engine in each scooter = ... × .... = ... N

So, the mass of student B and her scooter =

The maximum driving force is constant for a given engine.

Mass = ... kg
[4]

ii) Student A lends his scooter to student C. Student C weighs less than Student A. Describe how the acceleration of the scooter will differ when student C is riding it compared with when student A is riding it.

....

....
[1]
[Total 5 marks]

Exam Practice Tip

Momentum collision questions can look a bit scary, but if you keep a cool head and work through them step by step they're a doddle. Remember the total momentum of a collision is <u>zero</u> — that means the momentum before a collision is the same as the momentum after, so you can put them equal to each other.

Score

22

😐☐ 🙂☐ 😃☐

Section One — Forces and Their Effects

Tick the box that matches how confident you feel with the questions in each topic. This should help show you where you need to focus your revision.

Exam Tips

AQA Certificate Exam Stuff

1) You have to do two exams for the AQA Level 1/2 Certificate in Physics — Paper 1 and Paper 2 (ingenious).

2) Both exams are 1½ hours long, and worth 90 marks.

3) Both papers test your knowledge and understanding of Physics. No surprises there. But in Paper 2, there's more of a focus on experimental and investigative skills, like reading and drawing graphs, planning experiments, or evaluating conclusions.

There are a Few Golden Rules

1) **Always, always, always make sure you read the question properly.**
For example, if the question asks you to give your answer in mm, don't give it in cm.

2) **Look at the number of marks a question is worth.**
The number of marks gives you a pretty good clue of how much to write.
So if a question is worth four marks, make sure you write four decent points. And there's no point writing an essay for a question that's only worth one mark — it's just a waste of your time.

3) **Write your answers as clearly as you can.**
If the examiner can't read your answer you won't get any marks, even if it's right.

4) **Use specialist vocabulary.**
You know the words I mean — the sciencey ones, like rectification and fusion. Examiners love them.

> Obeying these Golden Rules will help you get as many marks as you can in the exam — but they're no use if you haven't learnt the stuff in the first place. So make sure you revise well and do as many practice questions as you can.

5) **Pay attention to the time.**
The amount of time you've got for each paper means you should spend about a minute per mark.
So if you're totally, hopelessly stuck on a question, just leave it and move on to the next one.
You can always go back to it at the end if you've got enough time.

6) **Show each step in your calculations.**
You're less likely to make a mistake if you write things out in steps. And even if your final answer's wrong, you'll probably pick up some marks if the examiner can see that your method is right.

You Need to Understand the Command Words

Command words are the words in a question that tell you what to do.
If you don't know what they mean, you might not be able to answer the questions properly.

Describe...   This means you need to recall facts or write about what something is like.

Explain...   You have to give reasons for something or say why or how something happens.

Give...   This means the same thing as 'Name...' or 'State...'.
You usually just have to give a short definition or an example of something.

Suggest...   You need to use your knowledge to work out the answer. It'll often be something you haven't been taught, but you should be able to use what you know to figure it out.

Calculate...   This means you'll have to use numbers from the question to work something out.
You'll probably have to get your calculator out.

Speed and Velocity

1 Most quantities can be put into one of two groups: scalars or vectors.

a) Describe the difference between a scalar and a vector quantity.

....

....
[2]

b) Which of the following is a vector? Place a tick in the appropriate box to indicate your answer.

☐ speed   ☐ distance   ☐ mass   ☐ force
[1]

c) Which of the following are scalar? Place a tick in the appropriate boxes to indicate your answer. Tick **two** boxes.

☐ 14 kg   ☐ 300 kN down   ☐ 24 m/s west   ☐ 1 hour 9 minutes
[1]

[Total 4 marks]

2 The diagram shows a cyclist at a particular point on his journey. He is travelling north.

speed: 7 m/s

a) The cyclist turns around and begins to travel south at the same speed.

Draw a ring around the correct answer to complete the sentence.

The cyclist's velocity │ has changed │
│ is the same │ .
│ is zero │
[1]

b) The cyclist travels 8000 metres north, then 224 metres south. It takes him 1440 seconds.

i) Calculate his total displacement, in metres, at the end of his journey.

Total displacement = ... m, direction = ...
[1]

ii) Calculate the average velocity, in metres per second, of the cyclist across his journey.

Average velocity = ... m/s, direction = ...
[2]

[Total 4 marks]

Score: ☐

8

Distance-Time Graphs

1 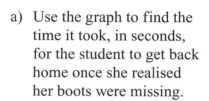 A student jogs to football training, then spends 78 seconds looking for her boots before she realises she has left them at home. She turns around and runs back to find them. A distance-time graph for her journey is shown.

a) Use the graph to find the time it took, in seconds, for the student to get back home once she realised her boots were missing.

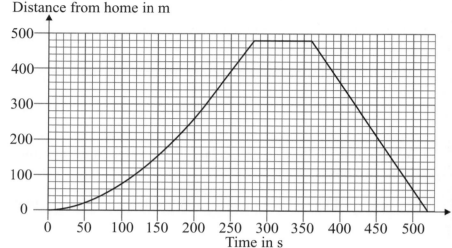

Time = ... s

[1]

b) Use the graph to calculate the student's average velocity towards her house, in metres per second, as she ran back home.

Average velocity = ... m/s

[3]

c) State whether the student ran to training at a steady speed. Explain how you know.

....

....

....

[2]

d) Using the graph, calculate how fast, in metres per second, the student was moving at 110 seconds.

Speed = ... m/s

[3]

[Total 9 marks]

Score:

9

Acceleration and Velocity-Time Graphs

1 A company produces battery-powered model cars.
Their latest model accelerates from rest to 20 m/s in 2.5 s.

a) Calculate the acceleration of the model car
during these 2.5 seconds.

Acceleration = ... m/s²
[2]

b) The car comes to a stop, then restarts with the same acceleration, travelling south. It maintains
this acceleration for 1.5 seconds. Calculate the velocity of the car to the south, in metres per
second, after this time has passed.

Velocity = ... m/s
[2]
[Total 4 marks]

2 A tractor is travelling east, accelerating at 0.5 m/s². After 8 seconds it reaches a speed of 5 m/s.

a) Calculate the eastward velocity, in metres per second, of the tractor before it started to accelerate.

Velocity = ... m/s
[2]

b) The tractor's eastward velocity increases from 5 m/s to 8 m/s in 12 seconds.
Calculate the tractor's average acceleration during this time, in metres per second squared.

Average acceleration = ... m/s²
[2]

c) The tractor is travelling east at 7 m/s when a dog runs into the road and
the tractor driver brakes. The tractor decelerates at 3.5 m/s².
Calculate how long, in seconds, the tractor takes to stop.

Time = ... s
[2]
[Total 6 marks]

Section One — Forces and Their Effects

3 The diagram shows a velocity-time graph for a car during a section of a journey.

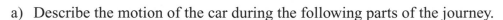

Velocity in m/s

Time in s

a) Describe the motion of the car during the following parts of the journey.

i) Between 40 and 60 seconds.

....

[1]

ii) Between 60 and 100 seconds.

....

[1]

b) Calculate the distance travelled, in metres, by the car between 40 and 60 seconds.

Distance travelled = ... m

[3]

c) Calculate the acceleration of the car, in metres per second squared, between 0 and 40 seconds.

Acceleration = ... m/s²

[3]

d) After 100 seconds, the car accelerates steadily for 40 seconds until it reaches a steady velocity of 30 m/s, which it maintains for 60 seconds. Complete the graph to show this motion.

[2]

[Total 10 marks]

Exam Practice Tip

Hopefully after that little lot you're now an expert in all things to do with acceleration. The two main things to remember are that the gradient of a velocity-time graph shows acceleration, and how to use the acceleration equation. You'll get given the equation in the exam, so you don't need learn it. Sweet.

Score

20

Resultant Forces

1   The diagram shows a cyclist sitting on a bicycle.

a) Explain why the weight of the cyclist does not cause him to fall through the bicycle saddle.

....

[1]

b) The cyclist travels east at a steady speed and then brakes.
This causes a *resultant force* and causes the cyclist to slow down.

What is meant by a *resultant force*?

....

....

[1]

[Total 2 marks]

2   A ball rolls to the west along the ground at a steady speed.

a) State whether there is a resultant force acting on the ball. Explain your answer.

....

....

[1]

b) A student picks the ball up and drops it. The ball accelerates towards the ground.
State whether there is a resultant force acting on the ball. Give a reason for you answer.

....

....

[1]

c) The ball hits the ground with a force of 5 N.
State the size of the force exerted on the ball by the floor. Explain your answer.

....

[1]

[Total 3 marks]

Score:

5

Combining Forces

1   The diagram shows the forces acting on a train as it pulls out of a station.

a) Calculate the resultant force, in newtons, acting on the train in the direction of travel.

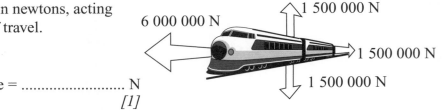

Force = ... N

[1]

b) The train passes a bird, which is flying due north with a thrust of 3 N. The train causes a crosswind with a force of 4 N blowing to the east as it passes. The scale diagram shows the forces acting on the bird.

What is the resultant force on the bird? State both its size, in newtons, and direction, as an angle from north.

Thrust is the force in the direction of travel.

1 cm = 1 N
drawn to scale

4 N

North   3 N

Force = ... N

Direction (angle from north) = ...°

[2]

[Total 3 marks]

2   The diagram shows two kites, A and B, flying in the sky. The forces acting on each kite at one instant are shown. The resultant force acting on kite B is zero.

a) Calculate the resultant force, in newtons, acting on kite A.

Force = ... N, direction = ...

[2]

b) Calculate the size, in newtons, of force *y*.

y = ... N

[1]

c) Calculate the size, in newtons, of force *x*.

x = ... N

[1]

[Total 4 marks]

Score:

7

Section One — Forces and Their Effects

Forces and Acceleration

1   The diagram shows a stationary camper van.
    The combined mass of the van and the driver is 2500 kg.

a) Use a word from the box to complete the following sentence.

unbalanced	changing	upwards	balanced

The forces on a stationary object are always ....

[1]

b) The camper van travels along a straight, level road at a constant speed of 90 kilometres per hour.

i) What is the resultant force on the camper van? Explain your answer.

    ....

    ....

[2]

ii) A headwind begins blowing with a force of 200 N, causing the
    van to slow down. Calculate the van's deceleration.

A headwind blows in the opposite direction to the van's motion.

                                            Deceleration = ... m/s²
[2]

c) The van travels at a steady speed before colliding with a stationary
    traffic cone with a mass of 10 kg. The traffic cone accelerates in the
    direction of the van's motion with an acceleration of 29 m/s².

i) Calculate the size of the force applied to the traffic cone by the van.

        $F = m \times a$
                                            Force = ... N
[2]

ii) State the size of the force applied by the cone to the van during the collision.

                                            Force = ... N
[1]

iii)Calculate the deceleration of the van during the collision.
    Assume all of the force applied by the cone to the van causes the deceleration.

                                            Deceleration = ... m/s²
[2]

[Total 10 marks]

fitted their scooters with the same engine. Student A and
[...] combined mass of 127.5 kg and a maximum acceleration of
[...]nt B has a maximum acceleration of 1.70 m/s² on her scooter.

[...]mbined mass, in kilograms, of student B and her scooter.

[...] force of the engine in each scooter = ... × .... = ... N

[...] of student B and her scooter =

The maximum driving force is constant for a given engine.

Mass = ... kg

[4]

b) Student A lends his scooter to student C. Student C weighs less than Student A.
Describe how the acceleration of the scooter will differ when student C
is riding it compared with when student A is riding it.

....

....

[1]

[Total 5 marks]

3   A 75 kg student runs west into a headwind of 20 N. Her overall acceleration is 0.5 m/s².

a) Calculate the resultant westward force, in newtons, on the student.

Resultant force = ... N

[2]

b) Calculate the size of the driving force, in newtons, that she is generating.
Assume no other forces are acting in the horizontal direction.

Driving force = ... N

[2]

c) The student exerts a force of 730 N on the road. State the force the road exerts on the student.
Explain your answer.

....

....

[1]

[Total 5 marks]

Score: ☐

20

☺☐ ☺☐ ☺☐

Momentum and Collisions

1 In a demolition derby, cars drive around an arena and crash into each other. Car A has a mass of 650 kg and a velocity of 15 m/s.

a) Calculate the momentum, in kilogram metres per second, of car A.

Momentum = ... kg m/s

[2]

b) Car A collides head-first with car B. Car B has a mass of 750 kg and travels at a velocity of 10.2 m/s before the collision. The two cars stick together. Calculate the combined velocity of the two cars immediately after the collision.

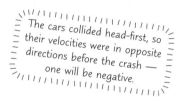

The cars collided head-first, so their velocities were in opposite directions before the crash — one will be negative.

Velocity = ... m/s

[3]

[Total 5 marks]

2 A skater with a mass of 60 kg is moving at 5 m/s. He skates past a bag and picks it up from the floor, causing him to slow down to 4.8 m/s.

Calculate the mass, in kilograms, of the bag. Assume there are no frictional forces.

initial momentum of skater = ... × ....

= ....

momentum of skater and bag = (........................ + mass$_{bag}$) × ....

Mass = ... kg

[Total 4 marks]

Exam Practice Tip

Momentum collision questions can look a bit scary, but if you keep a cool head and work through them step by step they're a doddle. Remember the total momentum of a collision is <u>zero</u> — that means the momentum before a collision is the same as the momentum after, so you can put them equal to each other.

Score
9

Momentum and Safety

1 A car with a mass of 1200 kg travels along the motorway at 30 m/s. It crashes into the central reservation barrier and is stopped in 1.2 seconds (after which its momentum is zero).

a) i) Calculate the momentum, in kilogram metres per second, of the car before the crash.

Momentum = ... kg m/s

[2]

ii) Calculate the size of the average force acting on the car during the crash.

Force = ... N

[2]

b) The car is fitted with air bags.

i) Explain how the air bags reduce the forces acting on the driver during the crash.

....

....

....

....

[3]

ii) State **one** car safety feature other than air bags that is also designed to reduce the forces acting on the driver during a crash.

....

[1]

c) Another car brakes with an average force of 4500 N as it passes the crash. It slows down from a velocity of 27 m/s to 12 m/s over 5 seconds. Calculate the mass, in kilograms, of this car.

Mass = ... kg

[3]

[Total 11 marks]

Score: ☐

11

Section One — Forces and Their Effects

 ☐   ☐   ☐

Frictional Force and Terminal Velocity

1 A truck moves forwards at a steady speed. The diagram shows the thrust (driving force) acting on it.

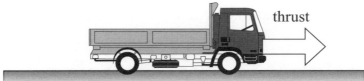

thrust

a) As the truck moves, it experiences resistance from friction. Add an arrow to the diagram to show the direction in which this resistance acts.

[1]

b) Draw a ring around the correct answer to complete the sentence.

The resistive force is
| the same size as |
| greater than |
| smaller than |
the thrust of the truck.

[1]

c) Describe how the speed of the truck affects the size of the resistance force it experiences.

....

....

[1]

[Total 3 marks]

2 A student investigates how the area of an object's parachute affects the forces acting on it as it falls. She attaches a parachute to a steel ball, drops it from a fixed height and times how long it takes to hit the ground. She repeats the experiment using parachutes of different sizes.

a) i) Explain why using the same steel ball throughout the experiment will improve the validity of the results.

parachute

stopwatch

steel ball

weight of ball and parachute

....

....

....

....

[2]

ii) Suggest and explain **one** other way in which the student can make sure the results are valid.

....

....

....

[2]

b) For each parachute, the steel ball initially accelerates before reaching terminal velocity.
Explain why a falling object reaches terminal velocity.

....

....

....

....

[3]

c) i) The table shows the velocity of the ball and parachute during one drop.
On the axes, draw a velocity-time graph for the ball during this fall.

Time in s	0	0.1	0.2	0.3	0.4	0.5	0.6
Velocity in m/s	0	0.75	1.0	1.1	1.1	1.1	1.1

velocity in m/s

1.25
1.00
0.75
0.50
0.25
0.00

0.0   0.1   0.2   0.3   0.4   0.5   0.6   time in s

[3]

ii) Use the graph to find the terminal velocity of the ball and parachute during this fall.

Terminal velocity = ... m/s

[1]

d) Parachutes are used by skydivers to help them land safely.
Describe how a parachute decreases a skydiver's terminal velocity.

....

....

....

[2]

[Total 13 marks]

Exam Practice Tip

Make sure you know what <u>valid</u>, <u>repeatable</u> and <u>reproducible</u> mean. An experiment is <u>repeatable</u> if you can repeat it and get the same results. It's reproducible if other scientists can do the experiment and get the same results. An experiment is valid if it's both repeatable and reproducible, and answers the original question.

Score

16

Stopping Distances

1   A car's stopping distance is equal to the driver's thinking distance plus the car's braking distance.

a) The thinking distance is the distance travelled during the driver's *reaction time*.

   i) What is meant by *reaction time*?

   ....
   [1]

   ii) Name **two** factors that can affect a driver's reaction time.

   1. ....

   2. ....
   [2]

b) The brakes of a car can be altered to increase the car's maximum braking force.
   Describe how this will this affect the car's braking distance.

   ....
   [1]

c) Other than the condition of a car's brakes, name **three** factors
   that can affect a car's braking distance.

   1. ....

   2. ....

   3. ....
   [3]

   [Total 7 marks]

2   A student makes the following statement about driving.

Heavy rain increases my stopping distance.

a) Explain **one** way in which heavy rain can increase a car's stopping distance.

   ....
   [1]

b) Suggest **one** way a driver could decrease their stopping distance if driving in heavy rain.

   ....
   [1]

   [Total 2 marks]

Score:

9

Section One — Forces and Their Effects

Weight, Mass and Gravity

1 A student conducted an experiment to measure gravitational field strength, *g*. He suspended an object with a mass of 2 kg from a newton-meter held in his hand, as shown in the diagram. He took multiple readings of the object's weight and calculated an average value of 19.6 N.

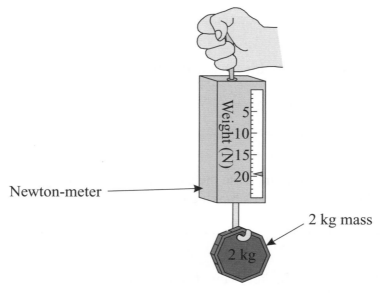

Newton-meter

2 kg mass

a) Calculate the gravitational field strength, in newtons per kilogram.

Gravitational field strength = ... N/kg

[2]

b) The student accidentally drops the 2 kg mass.
Assuming there is no air resistance, what is its acceleration?

Acceleration = ... m/s²

[1]

c) The gravitational field strength on the Moon is 1.6 N/kg. Calculate how much the 2 kg mass would weigh, in newtons, if the student repeated this experiment on the Moon.

Weight on the Moon = ... N

[2]

[Total 5 marks]

Score:

5

Work and Potential Energy

1 The diagram shows a cyclist riding on flat ground. She travels 1600 m and applies a steady force of 50 N as she cycles.

a) Use a word from the box to complete the sentence.

weight	mass	energy

When work is done, ... is transferred.

[1]

50 N

Work done = ... J

[2]

[Total 3 marks]

2 A climber climbs up a cliff. She uses a constant force to climb a vertical height of 120 m, doing 90 kJ of work in the process.

a) i) Calculate the size of the upwards force, in newtons, applied by the climber during her climb.

Force = ... N

[2]

ii) The climber does work against a downwards force.
What name is given to the force the climber does work against?

....

[1]

b) Calculate the mass, in kilograms, of the climber.

Mass = ... kg

[3]

[Total 6 marks]

Score:

9

Kinetic Energy

1   A roller coaster cart with a mass of 105 kg rolls along a horizontal track at 3.0 m/s.

3.0 m/s

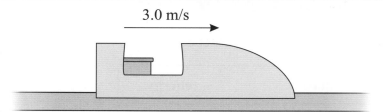

a)   Draw a ring around the correct answer to complete the sentence.

Kinetic energy is the energy an object has due to its

| movement. |
| height. |
| temperature. |

[1]

b)   Calculate the kinetic energy, in joules, of the cart.

Kinetic energy = ... J

[2]

c)   The cart is stopped at the end of roller coaster by applying brakes to the cart wheels.
Explain how the brakes slow down the cart.

....

....

....

[2]

[Total 5 marks]

2   A meteor has a kinetic energy of 67 500 kJ and travels
at a speed of 15 km/s as it enters the Earth's atmosphere.

a)   Calculate the mass, in kilograms, of the meteor as it enters the Earth's atmosphere.

Mass = ... kg

[2]

b)   Give the name of **one** type of energy that the kinetic energy of the meteor is transferred to as it
falls through the atmosphere. Explain what causes this energy transfer.

....

....

[2]

[Total 4 marks]

Section One — Forces and Their Effects

3   The diagram shows a truck and car travelling along a road. The car has a mass of
1500 kg. The truck has a mass of 3000 kg. Both vehicles travel at the same speed.

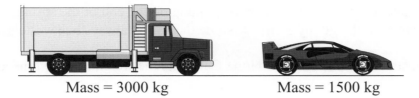

Mass = 3000 kg   Mass = 1500 kg

a) i) The kinetic energy of the car is 10 000 J.
        Calculate the kinetic energy, in joules, of the truck.

Kinetic energy = ... J

[1]

ii) The speed of the car doubles. Calculate the new kinetic energy, in joules, of the car.

Kinetic energy = ... J

[1]

b) i) The car accelerates until it has a kinetic energy of 432 000 J.
        Calculate the new velocity of the car.

Velocity = ... m/s

[2]

ii) The car has a maximum braking force of 7200 N.
    Calculate the minimum braking distance, in metres, of the car at this velocity.

Minimum braking distance = ... m

[3]

iii) The driver of the car increases his speed. Explain why it is important that the driver leaves
     more room between his car and the vehicle in front of him when his speed increases.

....

....

....

[2]

iv) Complete the sentence.

When the car's brakes are applied, the temperature of the brakes will ....

[1]

[Total 10 marks]

Score: [   ]

19

Section One — Forces and Their Effects

Forces and Elasticity

1 A 10 cm spring hangs from a hook. A mass is hung on the end of the spring, causing the spring to extend in length by 3 cm. Assume the spring always behaves elastically.

a) The spring constant of the spring is 2 N/m. Calculate the weight, in newtons, of the mass.

Weight = ... N

[2]

b) The mass is removed from the spring, causing the length of the spring to change. What is the length of the spring after the mass is removed? Explain your answer.

....

....

[1]

[Total 3 marks]

2 The diagram shows a gymnast on a trampoline. Springs around the edge hold the trampoline bed in place. The springs behave elastically.

trampoline bed   springs

a) When she is at the top of a bounce, the gymnast has **gravitational potential energy**. This is transferred to **kinetic energy** as she falls back down, but at the bottom of a bounce the kinetic energy is **zero**. Describe what happens to her kinetic energy.

....

....

[1]

b) The gymnast exerts a force of 600 N on the trampoline at the bottom of her bounce.

This force is split evenly across 30 identical springs that support the trampoline bed. Each individual trampoline spring extends by 10 cm at the bottom of her bounce. Calculate the spring constant of the springs, in newtons per metre.

Assume that only the springs extend.

Spring constant = ... N/m

[3]

[Total 4 marks]

3   A student investigated how a type of rope extends when a force is applied to it.

a) He suspended the piece of the rope vertically and hung different weights from the rope.
    He measured the length of the rope before it had any weights attached with a mm ruler
    (read at eye level). He measured the length of the rope when each weight was attached
    to it by holding the ruler in the same place.

    He found the extension of the rope for each weight by subtracting the original length
    of the rope from his length measurements.

    Suggest **two** ways in which he could change his method to improve the validity of his results.

    1. ....

    ....

    2. ....

    ....

    [2]

b) The student plotted this graph of force against extension using the results from his experiment.

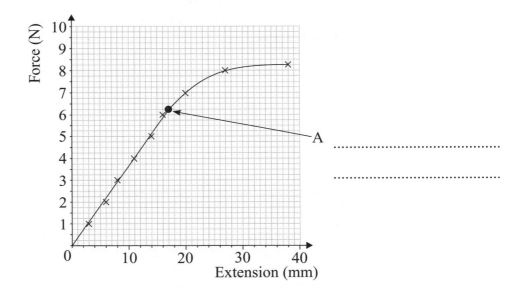

    ....

    ....

    i) Use a term from the box to label point A on the graph.

limit of proportionality	spring constant	elastic potential

    [1]

    ii) Give the range of forces within which the force on the rope is directly proportional to the
        extension of the rope.

    ....

    [1]

    [Total 4 marks]

    Score:

    11

Section One — Forces and Their Effects

Power

1 A student builds two battery powered model boats. Boat A has a 150 W electric motor. Boat B transfers 7800 J of energy every minute. Both boats only have one speed setting.

a) Calculate the power, in watts, of boat B.

Remember:
power = work done ÷ time

Power = ... W

[2]

b) i) Calculate the energy, in joules, that boat A transfers in 10 minutes.

Energy transferred = ... J

[2]

ii) The motor in boat A is replaced with a motor that has a power of 300 W, but is otherwise identical. State how you would expect this to change how often the boat's battery will need recharging. Explain your answer.

....

....

[3]

[Total 7 marks]

2 Three students each drag identical tyres as far as they can along the same bit of ground. The table shows the work done by each student while dragging their tyre and the time taken.

Student	Work done in J	Time taken in s
A	600	20
B	1000	40
C	50	5

Which of the students, A, B or C, was the most powerful?

Write your answer in the box.

Explain your answer.

....

....

[Total 2 marks]

Score:

9

Turning Forces and Centre of Mass

1 The diagram shows a rectangular board held at either end by two teachers. The board is 0.8 m long and is held so that the top and bottom edges of the board are horizontal.

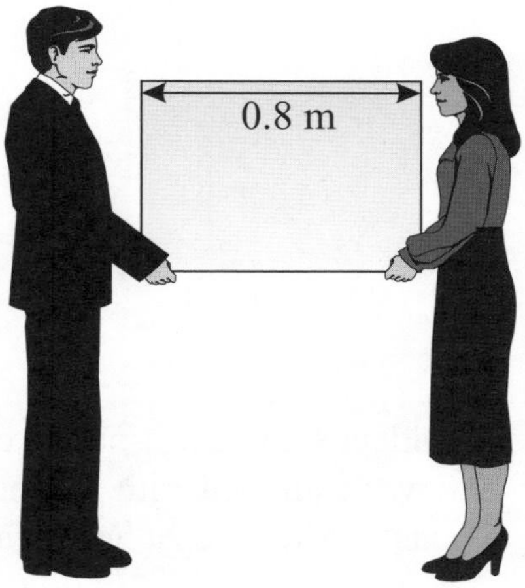

a) i) Describe what is meant by the *centre of mass* of an object.

....

....

....

....
[1]

0.8 m

ii) On the diagram, add a cross to mark the centre of mass of the blackboard.

[1]

b) One teacher applies an upwards force of 20 N to the end of the blackboard he is holding. The corner of the board held by the other teacher acts as a pivot. Calculate the moment, in newton-metres, of the force applied by the teacher.

Moment = ... Nm
[2]

[Total 4 marks]

2 A door has a horizontal door handle. To open the door, the handle needs to be rotated clockwise. The diagrams A, B, C and D show equal forces applied to the door handle at different positions and angles.

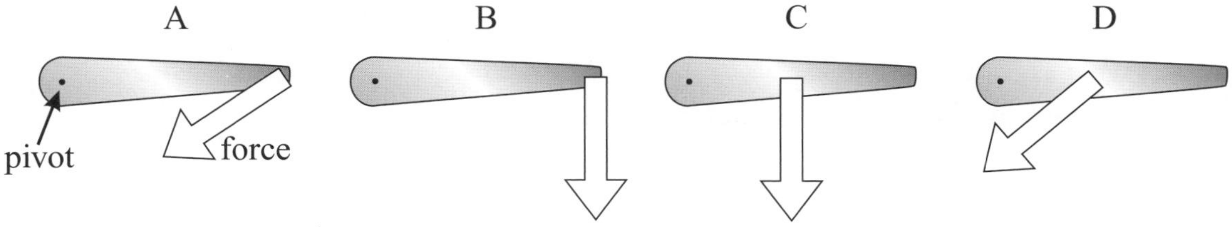

A   B   C   D

pivot   force

a) Which diagram shows the largest moment on the handle?

Write your answer in the box.

Explain your answer.

....

....
[2]

b) A force of 4.5 N is exerted vertically downwards on the door handle and creates a moment of 0.675 Nm about the pivot. Calculate the distance, in metres, between where the force is applied and the pivot.

Distance = ... m

[2]

[Total 4 marks]

3 A student wanted to find the centre of mass of an irregularly shaped piece of cardboard. She was equipped with a stand with a hook to hang the card from, a plumb line and a pencil. She made a hole near one edge of the card and hung it from the stand.

a) When she hung the card up, it swung slightly before it eventually came to rest. Explain why there were no moments acting on the card due to its weight when it was at rest.

....

....

[1]

b) i) Describe a method the student could have used to find the centre of mass of the cardboard.

....

....

....

....

....

[4]

ii) Suggest **one** way she could make her result more accurate.

....

....

[1]

[Total 6 marks]

Exam Practice Tip

The important thing to remember with moments is that it's the <u>perpendicular</u> distance to the pivot that counts, so the direction in which the force acts really matters. If two forces are the same size and act at the same distance from the pivot, they won't produce the same moment if they're at different angles to each other.

Score

14

Balanced Moments and Levers

1   The diagram shows three weights on a wooden plank, resting on a pivot.
    Weight A is 2 N and sits 20 cm to the left of the pivot. Weight B exerts
    an anticlockwise moment of 0.8 Nm. Assume the plank has no weight.

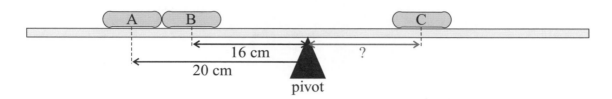

a) Calculate the anticlockwise moment about the pivot exerted by weight A.

Weight A anticlockwise moment = ... Nm

[2]

b) The system is currently balanced. Calculate the distance of weight C from the pivot
    if its weight is 8 N.

Distance = ... m

[3]

[Total 5 marks]

2   The diagram shows a wheelbarrow. Wheelbarrows
    are designed to make lifting heavy loads easier.

    Explain how a wheelbarrow reduces the
    amount of force needed to lift an object.

    *Hint — the wheel
    acts as a pivot...*

    handles

    weight of load and
    wheel barrow

    ....

    ....

    ....

    ....

[Total 2 marks]

Score: ☐

7

Moments, Stability and Pendulums

1   The diagram shows a cart being used to carry coal along a slope.
The centre of mass of the cart, when full, is shown.

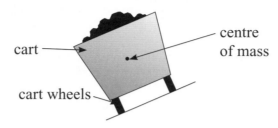

a) Explain why the cart **does not** topple over when on the slope.

....

....

Think about what force would cause a moment to make the cart topple.

[1]

b) Suggest **two** ways in which the cart could be altered to make it more stable.

1. ....

2. ....

[2]

[Total 3 marks]

2   Clock A has a pendulum that swings with a frequency of 0.625 Hz.
Clock B has a pendulum that swings with a time period of 1.25 seconds.

a) Calculate the time period of the pendulum in clock A.

Time period = ... s

[2]

b) Calculate the frequency of the pendulum in clock B.

Frequency = ... Hz

[2]

c) State how the time period of a pendulum can be altered.

....

....

[1]

[Total 5 marks]

Score: ☐

8

Circular Motion

1 The diagram shows a ball on a string being swung in a horizontal clockwise circle above a student's head. The ball travels at a constant speed along the path shown.

Ball

String

D

Circular path

a) The arrow labelled D shows the direction of the *centripetal force*.

   i) Describe what is meant by *centripetal force*.

   ....

   ....

   [1]

   ii) Name the force that provides the centripetal force on the ball.

   ....

   [1]

b) i) Explain why the ball is accelerating even though it is travelling at a constant speed.

   ....

   ....

   [1]

   ii) State which direction the ball accelerates in.

   ....

   [1]

c) The ball is replaced with a different ball with double the mass. How would this affect the centripetal force needed to keep the ball travelling along the same circular path at the same speed? Place a tick in the appropriate box to indicate your answer.

   ☐ The force needed would be greater.

   ☐ The force needed would be smaller.

   ☐ The force needed would be the same as for the original ball.

   [1]

d) Suggest **two** ways the student could decrease the force needed to keep the same ball moving with circular motion.

   1. ....

   2. ....

   [2]

   [Total 7 marks]

   Score: ☐

   7

Section One — Forces and Their Effects

Hydraulics

1 The diagram shows a simple hydraulic system. The system is made up of two pistons, D and E, that apply a pressure to a tank of liquid.

a) Use a word from the box to complete the sentence.

incompressible	elastic	compressed

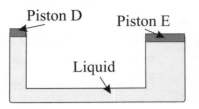
Piston D   Piston E

Liquid

Liquids are virtually ....
[1]

b) Hydraulic systems are used as *force multipliers* to lift heavy loads.
Explain how a hydraulic system works as a *force multiplier*.

....

....

....

....

....
[4]

c) Piston D has a cross-sectional area of 0.0025 m². Piston E has a cross sectional area of 0.09 m².
A force of 650 N is applied to piston D.

i) Calculate the pressure, in newtons per square metre, created in the liquid by piston D.

Pressure = ... N/m²
[2]

ii) What pressure is created in the liquid at piston E by the force at piston D?
Explain your answer.

....

....
[2]

iii) Calculate the force, in newtons, that is applied on piston E by this pressure.

Force = ... N
[2]

[Total 11 marks]

Score: ☐

11

 ☐ ☐ ☐

Wave Basics

1   Waves can be either transverse or longitudinal.

a) i) Complete the following sentence using a word from the box.

energy	amplitude	matter

Waves transfer .... from one place to another.

[1]

ii) Which of the following are waves **not** able to transfer?
Draw a ring around the correct answer.

matter   information   signals   energy

[1]

b) A student uses a spring to produce the two types of waves shown, type A and type B.

Type A   Type B

R: ....   S: ....

i) Label the features R and S of wave type A using words from the box.

compression	time period	rarefaction	diffraction

[2]

ii) Which wave type, A or B, is longitudinal?

Write the correct answer in the box.
Give a reason for your answer.

....

[1]

iii) Describe how the direction of energy transfer relates to
the direction of oscillation for transverse waves.

....

[1]

c) i) Name **one** example of a type A wave, other than a wave on a spring.

....

[1]

ii) Name **one** example of a type B wave, other than a wave on a spring.

....

[1]

[Total 8 marks]

A wave in a pond, travelling at 0.5 m/s, makes a floating ball move up and down twice every second.

a) What is the frequency, in hertz, of the wave?

wave speed 0.5 m/s

Frequency = ... Hz

[1]

b) The ball is on a crest of the wave. Calculate how far away, in metres, the next crest is from the ball.

Distance = ... m

[2]

c) Calculate the time period, in seconds, of the wave.

Time period = ... s

[1]

[Total 4 marks]

3   The diagram shows three electromagnetic waves, A, B and C.

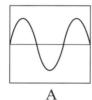 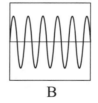 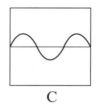

       A   B   C

a) i) What is meant by the *amplitude* of a wave? ....

....

[1]

ii) Which **two** waves have the same wavelength?
Write the correct answers in the boxes.   ☐   and   ☐

[1]

b) The diagram shows a graph of a water wave in a ripple tank.

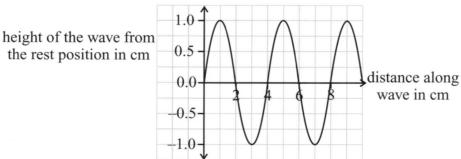

height of the wave from the rest position in cm

distance along wave in cm

Add to the diagram to show a water wave with the same amplitude but double the wavelength.

[1]

[Total 3 marks]

Score: ☐

15

Reflection of Waves

1   A ray of light hits a plane mirror at an angle of 20°, measured from the *normal*.

a) i) State the angle, measured from the normal, that the ray of light will be reflected at.

Angle of reflection = ... °

[1]

ii) State what is meant by the *normal* of a ray of light hitting a surface.

....

....

[1]

b) A student looks in the mirror at himself and sees an image formed from reflected light.

i) On the diagram, draw the paths of **two** rays of light to show how the student sees an image in the plane mirror of point A.

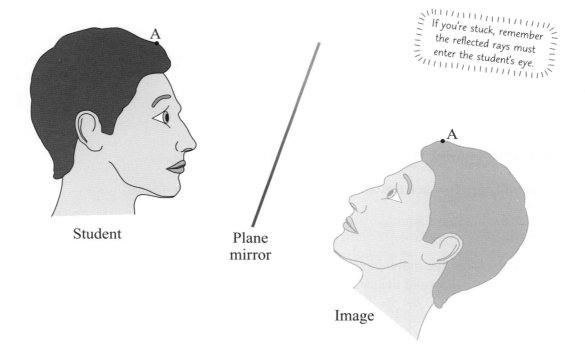

If you're stuck, remember the reflected rays must enter the student's eye.

Student

Plane mirror

A

Image

[2]

ii) Complete the following sentence using a word from the box.

virtual	real	enlarged	upside down

The image of the student is ....

[1]

[Total 5 marks]

Score: [   ]

5

Section Two — Waves

Diffraction and Interference

1 All waves can undergo *diffraction*.

a) Describe what is meant by *diffraction*.

....

....

[1]

b) i) The range of wavelengths of visible light is around 400-700 nm.
Explain why visible light does not appear to diffract through doorways.

....

....

[2]

ii) Explain why radio waves with a wavelength of 1 metre diffract noticeably as they pass through doorways.

Think about how their wavelength compares to the width of a doorway...

....

....

[1]

[Total 4 marks]

2 Two speakers in a room, speaker A and speaker B, produce sound with the same frequency and amplitude.

a) What will happen when the waves overlap at one point in the room?

....

[1]

b) The graphs show the individual sound waves produced by each speaker at one point in the room. The graphs are drawn to the same scale.

displacement   displacement   displacement

A   B

Draw a sketch graph on the blank set of axes to show the overall displacement of the air at this point in the room due to these sound waves.

[1]

[Total 2 marks]

Score: ☐

6

Refraction of Waves

1 The diagram shows the refraction of a ray of light travelling from material A into material B.

material B

material A

normal

a) Explain why the ray of light gets refracted.

....

....

....

[2]

b) Which material, A or B, is denser? Write your answer in the box.
Give a reason for your answer.

....

[1]

c) Describe how the path of the ray of light would be different if it had been travelling along the normal when incident on the boundary between the two materials.

....

[1]

[Total 4 marks]

2 The diagram shows a ray of red light entering a glass prism.

normal

air

incident ray   glass prism

a) Complete the diagram to show the ray passing through the prism and emerging from the other side. Label the angles of incidence, *i*, and refraction, *r*, for both boundaries.

[3]

b) Describe an experiment that you could use to measure *i* and *r* at both boundaries.

....

....

....

....

....

[4]

c) When a ray of white light travels through the prism it splits into separate colours. Name this effect.

....

[1]

[Total 8 marks]

Score:

12

Refractive Index

1 The diagram shows white light refracting at an air-glass boundary and separating into colours.

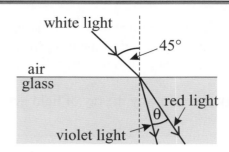

a) The speed of red light in air is 300 000 000 m/s.
The speed of red light in the glass is 200 000 000 m/s.
Calculate the refractive index of the glass for red light.

Refractive index = ...

[2]

b) The angle of refraction for red light in the diagram is 28.13°. The refractive index of the glass for violet light is 1.528. Calculate the angle θ shown in the diagram.

Angle of incidence for violet light = *i* = ...

$\sin r = \dfrac{\sin i}{n}$ (where *r* is the angle of refraction for violet light and *n* is the refractive index for violet light of the glass block).

$= \dfrac{\sin..........}{1.528} = $ .... $\Rightarrow r = $ ....

θ = ... − ....

=

θ = ... °

[4]

[Total 6 marks]

2 The diagram shows a ray of light shining into the water of a swimming pool. The refractive index of the water for the light is 1.3.

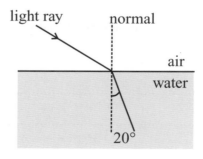

Calculate the angle of incidence, in degrees, of the light ray.

Angle of incidence = ... °

[Total 2 marks]

Exam Practice Tip

The equation for refractive index is a pretty nasty one, but with a bit of practice you'll be able to calculate any question on it the examiners can throw at you. Remember that if you need to find an angle, let's call it Fred, and you know what sin(Fred) is — just use the sin⁻¹ button on your calculator to find the angle.

Score

8

Total Internal Reflection

1 Optical fibres use *total internal reflection* to transmit light and information.

a) Describe what is meant by *total internal reflection*, and the circumstances in which it occurs.

....

....

....

[2]

b) State **one** situation in which optical fibres are used to transmit light and information.

....

[1]

[Total 3 marks]

2 A ray of blue light passes through the acrylic bottom of a boat into the water below it. The *critical angle (c)* of the acrylic-water boundary for blue light is 63.2°.

a) Describe what is meant by the term *critical angle*.

....

....

[1]

b) Complete the sentence using a phrase from the box.

equal to	greater than	less than

The angle of incidence of the ray of blue light at the boundary between

the acrylic and the water is .... the critical angle.

[1]

c) Another ray of blue light meets the acrylic-water boundary at an angle of incidence of 70°. Describe what will happen to this ray of light at the boundary.

....

[1]

d) The critical angle for light travelling from air into the acrylic is 41.8°. Calculate the refractive index of the acrylic in air.

Refractive index = ...

[2]

[Total 5 marks]

Score: ☐

8

 ☐   ☐   ☐

Electromagnetic Waves and Their Uses

1   The diagram shows six of the types of wave that make up the electromagnetic spectrum.

Radio waves	Microwaves	Infrared	Visible light	....	X-rays	Gamma rays

a) Complete the diagram by adding the missing type of electromagnetic wave.

[1]

b) State which type of electromagnetic wave has the lowest frequency.

....

[1]

c) Which of the following types of electromagnetic wave has the highest energy?
    Place a tick in the appropriate box to indicate your answer.

☐ radio waves   ☐ microwaves   ☐ gamma rays   ☐ visible light

[1]

d) Complete the sentence using a phrase from the box.

faster than	**at the same speed as**	**slower than**

Radio waves travel .... gamma rays in a vacuum.

[1]

e) State the entire range of wavelengths of the electromagnetic spectrum, from the shortest wavelength to the longest wavelength.

....

[1]

[Total 5 marks]

2   The diagram shows a radio transmitter that transmits long-wave radio waves.
    A mountain blocks the line-of-sight between the transmitter and a house.

radio transmitter

a) i) Explain why the mountain does not stop long-wave radio signals from reaching the house.

....

....

[2]

$10^4 \Rightarrow 10^{-15}$

ii) Give **one** other use of radio waves.

....
[1]

b) The house is fitted with a dish that receives satellite television signals.
State the type of electromagnetic radiation that is used to send satellite television signals.

....
[1]

c) The television in the house is mainly operated via remote control. What type of
electromagnetic wave do remote controls typically use to send signals to the television?
Draw a ring around the correct answer.

      infrared   **microwave**   **visible light**

[1]
[Total 5 marks]

3   A photographer marks his valuable equipment for security reasons.
Describe how electromagnetic radiation is used in security marking.

....

....

....
[Total 2 marks]

4   Mobile phones use electromagnetic waves for communication.

a) What type of electromagnetic wave do mobile phones use to receive and transmit calls?

....
[1]

b) Many types of mobile phone are able to connect with other devices
such as laptop computers over short distances via Bluetooth®.
Which type of electromagnetic wave is used by Bluetooth®?
Place a tick in the appropriate box to indicate your answer.

     ☐ radio waves   ☐ X-rays   ☐ microwaves   ☐ infrared

[1]

c) A student investigated the range of the Bluetooth® connection between his mobile phone and
his laptop. He tried to send a picture from his phone to his laptop while holding the phone at
different distances from the laptop.

i) Suggest and explain factors the student will have needed to keep the same in order to make sure his results were reliable.

....

....

....

....

[4]

ii) The student found the maximum distance from which he could successfully send a picture from his phone to his laptop was 7 metres.

The student left the laptop in his kitchen and went into a neighbouring room, shutting the door between the two rooms. He attempted to send the same picture from his phone to his laptop again. The distance between the phone and laptop was 5 metres.

Suggest **one** reason why the picture failed to reach his laptop.

....

[1]

d) Mobile phone screens emit visible light.
   Give **one** other example of visible light being used in communication.

....

[1]

[Total 8 marks]

5   The diagram shows electromagnetic radiation being used to sterilise a surgical instrument.

source of radiation

thick lead

a) What type of electromagnetic radiation is being used? Draw a ring around the correct answer.

microwaves   **X-rays**   **gamma rays**   **ultraviolet**

[1]

b) A similar process is used to treat fruit before it is exported to other countries. Suggest why this happens.

....

....

[2]

[Total 3 marks]

Score: ☐

23

Section Two — Waves

Dangers of Electromagnetic Waves

1 Overexposure to some types of electromagnetic radiation can be harmful.

a) Ultraviolet radiation is found in sunlight.
Give **two** health problems that can be caused by too much sun exposure.

1. ....

2. ....

[2]

b) Give **one** health problem that can be caused by exposure to gamma rays.

....

[1]
[Total 3 marks]

2 X-rays are a type of ionising radiation.

Explain why high doses of X-ray radiation are dangerous to living tissue.

....

....

....

[Total 2 marks]

3 Mobile phones use microwaves to transmit calls. A student
is worried about the possible dangers of using her mobile phone.

a) Suggest why she might be worried that excessive mobile phone use could be harmful.

....

[1]

b) Suggest why would it be more dangerous to use infrared
radiation instead of microwaves for mobile phone signals.

Think about the differences between waves with different frequencies.

....

....

[1]
[Total 2 marks]

Exam Practice Tip

As a rule of thumb, remember the higher the frequency of electromagnetic radiation,
the more dangerous it is. Radio waves are thought to be pretty harmless, but we have to be
really careful with the amount of ultraviolet waves, X-rays and gamma rays we're exposed to.

Score

7

X-Rays

1 X-rays can be used to treat cancer.
Explain why X-rays are used to treat cancer.

....

....

....

[Total 2 marks]

2 X-rays can be used to form medical images.

a) Complete the following sentence.

X-rays are a type of .... frequency electromagnetic radiation.

[1]

b) Describe how X-rays can be used to create medical X-ray 'photographs'.

....

....

....

absorbed by hard tissues, transmitted by rest

[3]

c) Name **one** other type of medical imaging that uses X-rays.

....

[1]

d) Name **one** medical condition X-ray images can be used to diagnose.

....

[1]

e) Radiographers take X-ray images of patients in hospitals.
Radiographers and patients need to limit their exposure to X-rays while an image is made.

i) State **one** precaution radiographers can take to minimise their exposure to X-rays.

....

[1]

ii) Describe **one** way in which the patient's radiation dose during a scan can be minimised.

....

[1]

[Total 8 marks]

Score:

10

Sound and Ultrasound

1   Sound waves are a type of longitudinal wave. The diagram shows three sound waves, A, B and C.

a)   Which sound wave is the loudest?
Write your answer in the box.

☐

[1]

b)   Which sound wave has the highest pitch?
Write your answer in the box.

☐

[1]

A

B

C

All graphs are
drawn to the
same scale.

c)   Another sound wave, D, has such a high frequency that it cannot be heard.

i)   What is the audible range of sound frequencies for a person with good hearing?

....
[1]

ii) Sound with a frequency above this range is called ultrasound.
Give **one** way ultrasound waves can generated.

....
[1]

d)   Sound waves cannot travel in a vacuum. Explain why.

....

....
[1]

[Total 5 marks]

2   A student sings in her school choir. She practises in both an empty drama hall and on the school playing field. She notices that her voice echoes only when she sings in the drama hall.

a)   Explain why her voice echoes in the drama hall but not on the playing field.

....

....
[1]

b)   The drama hall is fitted with sound insulation so that sound cannot pass through the walls and windows. A friend notices that if the door is open, she can still hear the student singing when she stands outside the drama hall and round the corner from the door. Explain why.

....
[1]

[Total 2 marks]

Score: ☐

7

More on Ultrasound

1 Ultrasound is the name given to sound waves above the pitch of human hearing. It is used for medical imaging.

a) Describe how an ultrasound image is formed during a medical scan.

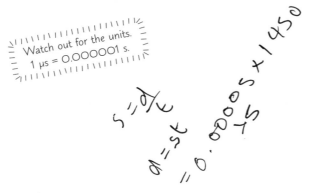

....

....

....

[2]

b) State and describe **one** other medical use of ultrasound.

....

....

[2]

[Total 4 marks]

2 A layer of fat between a layer of skin and a layer of muscle is examined using ultrasound.

An ultrasound pulse is emitted by a device which also acts as a receiver.
The oscilloscope trace shows two reflections of the ultrasound pulse detected by the receiver.

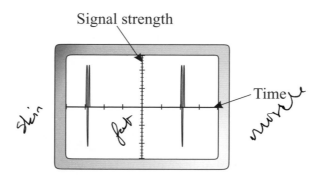

Signal strength

Time

The first pulse detected by the receiver is reflected at the boundary between the skin and the fat, and the second is reflected at the boundary between the fat and the muscle. Each time division on the oscilloscope represents 5 μs and the speed of sound through fat is 1450 m/s.

Calculate the thickness, in millimetres, of the layer of fat.

Watch out for the units.
1 μs = 0.000001 s.

Thickness = ... mm

[Total 2 marks]

Score:

6

Lenses and Magnification

1 A student uses a magnifying glass to focus rays of parallel light to form an image.

a) What type of lens can be used as a magnifying glass? Draw a ring around the correct answer.

      converging   **diverging**   **concave**

[1]

b) Describe how a magnifying glass brings parallel rays of light together.

....

....

[1]

c) What is meant by the *principal focus* of a lens?

....

....

[1]

d) The magnifying glass has a *focal length* of 10 cm. What is meant by the term *focal length*?

....

....

[1]

e) Describe the difference between a real and a virtual image.

....

....

....

[2]

f) Give **two** ways in which the image of an object produced
by a lens may appear different from the object itself.

1. ....

2. ....

[2]

g) The student uses the magnifying glass to look at a mark on a table. He sees an image
of the mark that is 10 millimetres long. The magnification produced by the lens is 2.5.
Calculate the length, in millimetres, of the mark.

$$\frac{image\ h}{o\ h} = \frac{10}{x} = 2.5$$

Length = ... mm

[2]

[Total 10 marks]

$$\frac{10}{2.5} = 4 = x$$

Score: ☐

10

Section Two — Waves

Converging Lenses

1   Diagram 1 shows three parallel rays of light arriving at a converging lens.

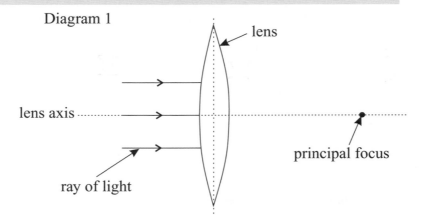

Diagram 1

a)   Which of the following does **not** have its direction changed by a converging lens.
     Place a tick in the appropriate box to indicate your answer.

☐ Any ray parallel to the axis.   ☐ Any ray passing through the centre of the lens.

☐ Any ray passing through
  the principal focus.   ☐ Any ray passing through the lens at
                                   an angle to the axis other than 0°.

[1]

b)   Add to the diagram to show the path of the rays
     passing through the lens and being brought to a focus.

[2]

c)   The lens is used to look at an object. Diagram 2 shows the object, the lens, and the points one
     focal length, F, and two focal lengths, 2F, away from the centre of the lens on the lens axis.
     The diagram is drawn to scale.

     Complete the diagram to show the image formed by the lens. Include **two** rays of light travelling
     from the object to the image in your answer.

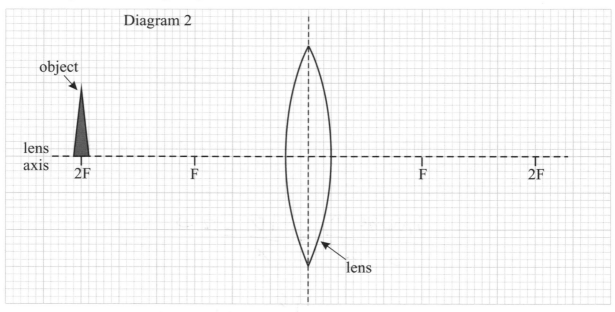

[4]

[Total 7 marks]

2 A student wants to find the focal length of a lens.
She has a screen, a ruler, and the lens.

a) Describe an experiment she could carry out to find the focal length of the lens.

....

....

....

....

....

....

....

[3]

b) The student measures the focal length of the lens five times, and calculates an average of 8.43 cm.
Suggest why the student measured the focal length five times.

....

....

[1]

[Total 4 marks]

3 A biologist places a magnifying glass 5 cm away from a ladybird.
The focal length of the lens is 7 cm.

a) Magnifying glasses use converging lenses. Which of the following is another name for a
converging lens? Draw a ring around your answer.

convex   **diverging**   **concave**   *[1]*

b) Describe the image produced by the lens in terms of its size, image type and orientation.

....

....

[3]

c) The ladybird moves until it is 10 cm away from the lens.
Describe how the image produced by the lens changes.

....

....

[2]

[Total 6 marks]

Score:

17

Section Two — Waves

Diverging Lenses

1 Diagram 1 shows three parallel rays arriving at a diverging lens.

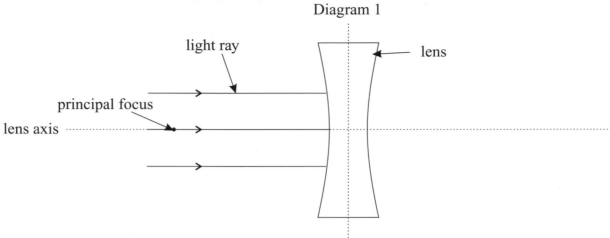

Diagram 1

a) State another name for a diverging lens.

....

[1]

b) State the properties of an image produced by a diverging lens.

....

....

....

[3]

c) Add to Diagram 1 to show how the parallel rays of light shown are refracted by the lens.

[1]

d) The lens is used to look at an object. Diagram 2 shows the object, the lens, and the points one focal length, F, and two focal lengths, 2F, away from the centre of the lens on the lens axis. The diagram is drawn to scale.

Complete the diagram to show the image formed by the lens. Include **two** rays of light travelling from the object to the image in your answer.

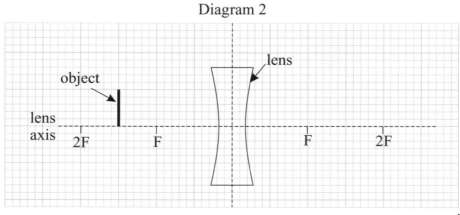

Diagram 2

[3]

[Total 8 marks]

Score:

8

Power and the Lens Equation

1   The greater the power of a lens, the more it refracts the light that enters it.

a)   A converging lens has a focal length of 40 cm. Calculate the power, in dioptres, of the lens.

Power = ... D

[1]

b)   The diagram shows a lens with a power of 20 D placed near an object. It creates an image 30 cm from the centre of the lens.

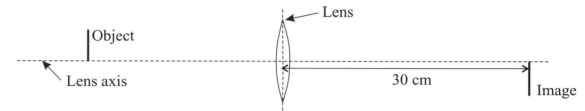

Calculate the distance, in centimetres, between the centre of the lens and the object.

$$\text{Power} = \frac{1}{\text{.....................}} = \text{.................}$$

The lens equation is: $\dfrac{1}{\text{.....................}} = \dfrac{1}{\text{.....................}} + \dfrac{1}{\text{.....................}}$

So...

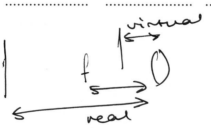

Object distance = ... cm

[3]

c)   The object is moved and placed 2 cm away from a convex lens with a 4 cm focal length. Is the image produced real or virtual? Explain your answer.

....

[1]

d)   Lenses A and B are both convex. Lens A has a focal length of 9 cm and lens B has a focal length of 7 cm. Suggest **two** possible differences between the lenses that could cause this difference in focal length.

1. ....

....

2. ....

....

[2]

[Total 7 marks]

Score: ☐

7

 ☐   ☐   ☐

Section Two — Waves

The Eye

1   The diagram shows the structure of the human eye.

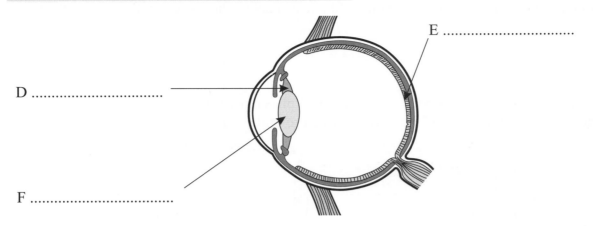

E ....

D ....

F ....

a)   Use the words in the box to label the parts of the eye, D, E and F, in the diagram.

| lens | retina | suspensory ligaments | ciliary muscle | cornea |

[3]

b) i) Which part of the eye controls how much light enters the eye?
        Draw a ring around the correct answer.

        the ciliary muscle   **the lens**   **the iris**   **the retina**

[1]

   ii) On which part of the eye is an image formed?

       ....

[1]

   iii)Describe the properties and function of the cornea.

       ....

       ....

       ....

       ....

[3]

[Total 8 marks]

2   The human eye typically has a *near point* of 25 cm.

   a)   What is meant by the term *near point*?

        ....

        ....

[1]

b) Where is the far point of a typical human eye?

....

[1]

c) Complete the sentence using a phrase from the box.

The distance between the near point and the far point is known as the .... of vision.

focal length	limit	range	power

[1]

d) Explain how the eye is able to focus on close and distant objects.

thicker

...close —> ciliary muscles contracts...
...far —> relax —> less thick...
...so the eye can focus on different...
....
....

[4]

[Total 7 marks]

3 A student made the following statement.

The human eye works like a camera.

Explain how the structure and operation of the human eye is similar to a camera.

...iris + apperture control light in...
eye focus using ciliary muscle - camera...
....
...both make real INVERTED image by...
focusing light thru lense...
....
...light falls on retina like CCD or film in cam...

[Total 4 marks]

Score:

19

Correcting Vision

1 The ray diagram shows what happens when a student with a vision problem attempts to focus on a distant object.

eye

object ●

a) What is the name given to the student's vision problem?

....

[1]

b) Name **two** possible causes of this vision problem.

1. ....

2. ....

[2]

c) The student decides he does not want to wear glasses and researches *laser* eye surgery.

i) What is a *laser*?

....

....

[1]

ii) Describe how laser eye surgery could be used to correct the student's vision.

....

....

[1]

[Total 5 marks]

2 People with long sight are unable to focus on close objects.

a) What type of lens is used to correct long sight?

....

[1]

b) Explain how wearing glasses with these lenses can correct long sight.

focus behind – not contract strongly enough

converge rays bf entering eye

focus image to on retina

+form

[4]

[Total 5 marks]

Score: ☐

10

The Doppler Effect and Red-Shift

1 The diagram shows an ambulance driving at high speed with its siren sounding. The ambulance passes a stationary pedestrian.

siren

AMBULANCE

a) i) How does the pitch of the siren change as the ambulance passes the pedestrian?

↑ wavelengths ↓ frequency ↓ pitch

[1]

ii) What is the name of this effect?

[1]

b) Explain why the pitch of the siren changes in this way.

[4]

c) Name **one** other type of wave that can show this effect.

[1]

[Total 7 marks]

2 Astronomers study the *red-shift* of light from distant galaxies.

a) What is meant by the term *red-shift*?

increased OBSERVED wavelength

[1]

b) How does the red-shift of light from distant galaxies suggest the universe is expanding?

(increased wavelength) moving away more distant → moving faster

[2]

[Total 3 marks]

Score:

10

We love breadth

Section Two — Waves

The Big Bang

1   The *Big Bang theory* and the Steady State theory are
     both theories explaining the origin of the universe.

a) i) What is the *Big Bang theory*?

    ....

    ....

    [1]

    (handwritten: small point ~ 2 explosded + expanding)

    ii) The red-shift of distant galaxies could be explained by both theories.

        Explain how the evidence from red-shift supports the *Big Bang theory*.

        ....

        [1]

b) The *Big Bang theory* is the only theory that provides an
    explanation for *cosmic microwave background radiation*.

    i) What is *cosmic microwave background radiation*?

        ....

        ....

        [2]

        (handwritten: low frequency electro magnetic radi...)

    ii) How does the *Big Bang theory* explain *cosmic microwave background radiation*?

        ....

        ....

        [1]

c) A student makes the following statement about the *Big Bang theory*.

        All the evidence supports the Big
        Bang theory, so it must be true.

        Hint — is any theory ever
        definitely, <u>definitely</u> true?

    Explain why the student is incorrect.

    ....

    ....

    [1]

    [Total 6 marks]

    Score: ____

    6

Kinetic Theory and Changes of State

1 Substances can exist in different states of matter.

a) i) Use words from the box to label the diagrams X and Y of the particles in two states of matter.

solid
liquid
gas

X ....

Y ....

[2]

ii) Which state of matter has the **highest** average energy per particle for a given substance?

....

[1]

b) A beaker of solid wax is heated constantly for 20 minutes.

i) Explain how heating a solid can cause it to turn into a liquid.

....

....

....

[2]

ii) The graph shows how the temperature of the wax changes over time whilst being heated.

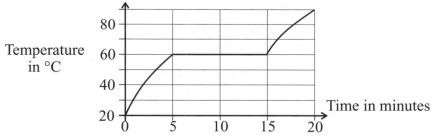

Describe and explain the shape of the graph between 5 and 15 minutes.

....

....

....

[3]

iii) After how many minutes of heating was the wax completely liquid?

Time = ... min

[1]

iv) Suggest why the melting point of another sample of the same wax might be different.

....

[1]

[Total 10 marks]

Score: ☐

10

Specific Heat Capacity

1   The table shows a list of materials and their *specific heat capacities*.

Material	Specific heat capacity (J/kg°C)
Concrete	880
Oil	2000
Mercury	139
Water	4200
Copper	380

a) i)   What is meant by the *specific heat capacity* of a substance?

....

....

[1]

ii) Which material will release the least heat energy when 1 kg of it cools by 10 °C?
Give a reason for your answer.

....

....

[2]

b)   A piece of copper is heated to 90 °C and lowered into a container of water.
The copper transfers 3040 J of energy to the water before it is removed from the container.
The temperature of the copper after it is removed is 50 °C.

Calculate the mass, in kilograms, of the piece of copper.

Mass = ... kg

[2]

c)   A student heats 27.2 kg of mercury and 2 kg of water to 70 °C. The student allows both liquids to
cool to 20 °C. Calculate the difference in energy released, in joules, by the two substances.

Difference in energy released = ... J

[3]

[Total 8 marks]

Exam Practice Tip

Specific heat capacities aren't too complicated, but make sure you're careful with your units.
If you're given a mass in grams or an amount of energy in kilojoules, make sure you change them
to be in kilograms and joules before you substitute them into the specific heat capacity equation.

Score

8

Section Three — Heating Processes

Specific Latent Heat

1 The table shows the melting point, boiling point and specific latent heat of fusion and vaporisation for three different materials.

Substance	Melting point (°C)	Specific latent heat of fusion (kJ/kg)	Boiling point (°C)	Specific latent heat of vaporisation (kJ/kg)
Water	0	334	100	2260
Copper	1083	205	2595	4730
Lead	327	23	1750	859

a) i) What is meant by the *specific latent heat of fusion* of a substance?

....

[1]

ii) Which material requires the most energy per kilogram to turn from a liquid to a gas?
Draw a ring around the correct answer.

       water   **copper**   **lead**

Give a reason for your answer.

....

[2]

b) A heat source supplies 184 kJ of thermal energy over 10 minutes. Use the table to calculate the maximum mass of lead, in kilograms, that the heat source could be used to melt in 10 minutes.

Mass of lead melted = ... kg

[2]

c) i) A student put 30 g of ice in a beaker of water and waited for all of the ice to melt. The ice was at an initial temperature of 0 °C. Calculate the energy transferred, in joules, to the ice by the water as the ice melted.

Energy transferred = ... J

[2]

ii) The student emptied the beaker and refilled it with 0.15 kg of water. He then heated the beaker using a heat source that supplied energy to the water at a constant rate of 250 J per second. When the water reached 100 °C he started a stopwatch. How long, in seconds, would the student need to continue heating the beaker in order for all of the water to boil away?

Time = ... s

[3]

[Total 10 marks]

Score:

10

Heat Radiation

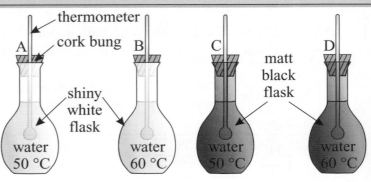

1   Four identically sized flasks made from the same material are filled with water of different temperatures. Each flask is sealed and a thermometer placed inside, as shown in the diagram.

a)   Flask C emits more radiation than flask B, even though the water inside it is cooler. Suggest why.

....

....
[2]

b)   i)   The temperature in the room is 22 °C. State whether the flasks will absorb radiation from the surroundings. Give a reason for your answer.

....

....
[2]

ii)   State which flask, A or B, you would expect to lose the most heat by radiation in 10 seconds. Explain your answer.

....

....
[1]

c)   Flasks A and B are placed in an ice bath and the temperature inside each flask drops rapidly. A student notices that the temperature inside flask B drops more quickly. Suggest why.

....
[1]

d)   In another experiment, flasks A and C are filled with cold water. Both flasks are heated equally with an infrared heater. After 5 minutes, the temperature of the water in flask C was higher than in flask A. Suggest why.

....

....
[2]

[Total 8 marks]

2   A solar panel consists of water pipes underneath a panel designed to absorb energy from the Sun.

Suggest a suitable colour and texture for the panel material. Give a reason for your answer.

....   Score:

....   **10**

[Total 2 marks]

Conduction and Convection

1 Heat can be transferred in different ways.

a) In what type of substance can convection **not** take place? Draw a ring around the correct answer.

        solid   **liquid**   **gas**

[1]

b) Explain why solids are better conductors of heat than liquids.

....

....

[2]

c) A student is carrying out an experiment in class to demonstrate convection. She fills a rectangular glass tube with water and heats one of the bottom corners, as shown.

i) Draw **two** arrows on the diagram to show the movement of the water within the tube.

[1]

ii) Explain, in terms of particles, why the water in the tube moves in the way that you have shown in part i).

....

....

....

....

[3]

d) During the experiment, both the metal Bunsen Burner and the plastic mat heats up.

i) Explain, in terms of particle movement, how heat is transferred through the mat by conduction.

....

....

[2]

ii) The Bunsen burner heats up much more quickly than the mat during the experiment. Explain why metals are good conductors.

....

....

[1]

[Total 10 marks]

Score:

10

Condensation and Evaporation

1 Condensation often forms on car windows on cold mornings.

Explain why water vapour condenses when it comes into contact with car windows.

....

....

....

....

Think about how energy is transferred between the water vapour and the car window.

[Total 3 marks]

2 A student goes for a long run and notices that his body begins to sweat a lot.

a) *In this question you will be assessed on the quality of your English, the organisation of your ideas and your use of appropriate specialist vocabulary.*

Explain how sweating helps to regulate the student's body temperature when he exercises. In your explanation, comment on the movement of particles in the sweat.

The student gets hot while doing exercise and sweats. Some particles in the liquid sweat...

....

....

....

....

....

[6]

b) Use a words or phrase from the box to complete the sentence.

more humid	colder	less humid

The student's sweat will dry more quickly when

his surroundings are ....

[1]

c) The student notices that when the wind is blowing, his sweat dries much more quickly. Explain why this happens.

....

....

....

[2]

[Total 9 marks]

Score:

12

Rate of Heat Transfer and Expansion

1 Animals have ways to help them control the amount of heat they transfer.

a) The diagram shows two different types of fox, A and B. Which fox you would expect to live in a hot environment? Explain your answer.

....

....

....

Fox A   Fox B

[2]

b) Leading up to winter, foxes grow longer fur. Explain how this can help reduce the amount of heat energy transferred from the fox to its surroundings.

....

Think about what types of heat transfer long fur might reduce...

....

[2]

[Total 4 marks]

2 A computer processor is connected to a metal cooling fin to prevent it overheating.

a) State and explain how features of the cooling fin make it effective at cooling the processor down.

....

....

....

....

[4]

b) A conducting thermal gel is used to fill air gaps between the processor and the fin. Suggest why.

....

[1]

c) i) When the processor temperature rises above 60 °C, a thermostat containing a bimetallic strip triggers a cooling fan, as shown in the diagram. Explain why the strip bends when the temperature rises.

metals joined together

metal 1 →   ← heat →

metal 2 →

....

....

[2]

ii) Give **one** example of where the effect you explained in part i) can become a hazard.

....

[1] Score: ☐

[Total 8 marks]

12

Section Three — Heating Processes

Energy Transfer and Efficiency

1   A torch emits 8 J of light energy, 11.5 J of heat energy and 0.5 J of sound energy every second. You may assume that these are the only energy outputs of the torch.

a) i) Which of the following sentences are **false**? Tick **two** boxes.

☐ Energy can be created.   ☐ Energy cannot be transferred.

☐ Energy cannot be destroyed.   ☐ Energy can be stored.

☐ Energy can be wasted.   ☐ Energy can be dissipated.

[2]

ii) How much energy is supplied to the torch each second? Explain your answer.

....

....

....

[2]

b) i) What is the useful output energy, in joules, of the torch?

Useful output energy = ... J

[1]

ii) Calculate the efficiency of the torch.

The input energy is the total amount of energy transferred.

Efficiency = ...

[2]

c) i) The battery of the torch is rechargeable. The battery charger operates at 20 W and is 75% efficient. Calculate the power, in watts, that is supplied to the battery while it is charging.

Power = ... W

[2]

ii) Describe what will eventually happen to all of the energy wasted by the battery charger.

....

[1]

[Total 10 marks]

Score: ☐

10

☺ ☐   ☺ ☐   ☺ ☐

Sankey Diagrams

1   A winch uses a cable and a hook to lift a weight by winding the cable around a drum.

a)   The manufacturer of a toy winch creates a Sankey diagram to show the energy transfers involved when the winch is in operation.

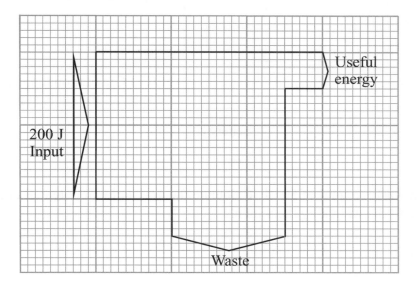

i)   Calculate the energy, in joules, represented by each small square.

1 square = ... J
[1]

ii) Calculate how much energy, in joules, is transferred usefully by the toy winch for every 200 J of energy supplied.

Useful energy transferred = ... J
[1]

iii) Four fifths of the energy wasted is wasted as heat energy and one fifth as sound.
     Use the grid below to draw a Sankey diagram for the toy winch to show this information.

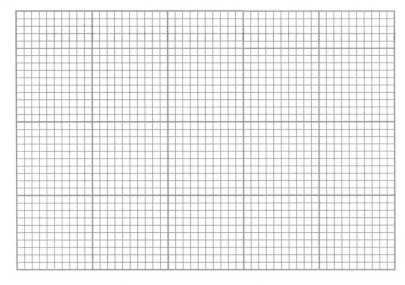

[3]

b) The diagram shows a Sankey diagram for a real winch lifting a weight.

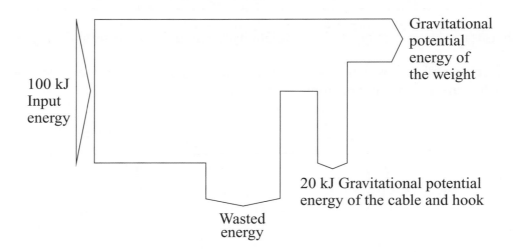

100 kJ
Input
energy

Gravitational
potential
energy of
the weight

20 kJ Gravitational potential
energy of the cable and hook

Wasted
energy

i) The winch wastes a total of 50 kJ. Calculate the gravitational potential energy, in kilojoules, transferred to the weight by the winch.

Gravitational potential energy transferred to the weight = ... kJ
[1]

ii) The weight is released and falls to the ground. 1.5 kJ of energy is transferred into heat and sound energy during the fall due to the air resistance on the weight. Sketch and label a Sankey diagram to show the energy transfers that take place during the weight's fall.

[3]

[Total 9 marks]

Energy Efficiency in the Home

1 A homeowner wants to reduce heat loss through the walls, windows and roof of her house.

a) The outer walls of the house are made up of two layers of bricks separated by an air cavity.

    i) Which type of energy transfer does having an air gap in the wall help to reduce?
Draw a ring around the correct answer.

        conduction   **radiation**   **convection**

        [1]

    ii) The homeowner is considering having cavity wall insulation installed.
State **one** type of energy transfer this will help to reduce. Explain your answer.

        ....

        ....

        [1]

b) The table shows the costs and savings of double glazing and insulated window shutters. A salesman tells the homeowner that insulated shutters are more cost-effective than double glazing.

	Double glazing	Insulated window shutters
Initial cost	£3000	£1200
Annual saving	£60	£20
Payback time	50 years	....

    i) Complete the table by calculating the payback time, in years, for insulated shutters.

        [1]

    ii) Explain **one** way in which the salesman's advice may not be correct.

        ....

        ....

        [1]

    iii) Suggest **one** other method of reducing loss from windows.
Explain how your suggestion will help reduce heat loss.

        ....

        ....

        [2]

c) The homeowner is choosing between two brands of loft insulation material.
Brand A has a U-value of 0.15 W/m²K. Brand B has a U-value of 0.2 W/m²K.

If both brands are the same price, which brand, A or B, should the homeowner should buy?
Explain your answer

        lowr = better

    ....

        [1]

[Total 7 marks]

Score: ☐

Current and Potential Difference

1   A student sets up a simple circuit containing a 3 V rechargeable battery, a length of wire and an ammeter. The battery supplies a potential difference which produces a reading of 5 A on the ammeter for 20 minutes before the battery needs recharging.

a) i) Calculate how much charge, in coulombs, will pass through the circuit before the battery needs recharging.

$$I = \frac{Q}{t}$$

Charge = ... C

[2]

ii) Calculate the energy transferred by the battery, in joules, before it needs recharging.

$$V = \frac{E}{Q}$$

Energy transferred = ... J

[2]

b) Electrical charge flows easily through the wire used in the circuit.

i) What name is given to a flow of electric charge?

....

[1]

ii) Suggest **one** material that the wire could be made from. Give a reason for your answer.

....

....

[2]

c) The student fully recharges the battery and replaces the wire in the circuit with a different one. The circuit has a much higher resistance than the first circuit.

Explain how this will affect the reading on the ammeter when a current flows through the circuit.

....

[1]

[Total 8 marks]

Exam Practice Tip

There are a couple of formulas to you need to be able to use in this topic — make sure you can pop them in a formula triangle and rearrange them. Don't forget that potential difference and voltage mean the same thing... they'll use potential difference in the exam, but you'll get the marks for using either. Lovely.

Score

8

Circuits — The Basics

1 Circuit symbols can be used to represent components in circuit diagrams.

a) Which of the following circuit symbols represents a fuse? Draw a ring around the correct answer.

[1]

b) A student wants to produce a graph of current against potential difference for component X.
An incomplete diagram of the circuit he is going to use is shown below.

B ....

X

C ....

i) Use words from the box to label components **B** and **C** on the diagram.

| variable resistor | battery | cell | thermistor |

[2]

ii) Add the letters **V** and **A** to the diagram to show where the ammeter and voltmeter should be.

[1]

c) *In this question you will be assessed on the quality of your English, the organisation of your ideas and your use of appropriate specialist vocabulary.*

Describe a method the student could use to obtain a good set of data to produce his graph.

the fix resistance
measure A using A & V using V
change (
Repeat 3 times for each

[6]

d) Suggest **one** quantity that the student could calculate from the graph he draws.

....

[1]

[Total 11 marks]

Score:

11

Section Four — Electricity

Resistance and V = I × R

1 The diagram shows current-potential difference (*I-V*)
graphs for four components at a constant temperature.

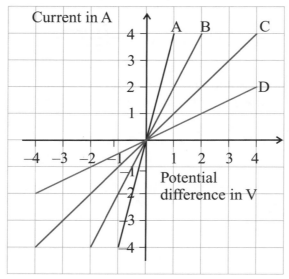

a) Draw a ring around the correct answer to complete the sentence.

All four components are types of

resistor.

filament lamp.

diode.

[1]

b) Which component, A, B, C or D, has the highest resistance?

Write your answer in the box.

[1]

c) i) Calculate the resistance, in ohms, of component B.

$$V = IR$$

Resistance = ... Ω

[3]

ii) The resistance of component B is tested at different temperatures.
At 30 °C, it has a resistance of 0.75 Ω when the potential difference across it is 15 V.

Calculate the current, in amps, through the component.

Current = ... A

[2]

[Total 7 marks]

2 A student tries to identify two components using a standard test circuit. The table below shows his sets of readings of current and potential difference for the two components.

Potential difference (V)	−4.0	−3.0	−2.0	−1.0	0.0	1.0	2.0	3.0	4.0
Component A current (A)	0.0	0.0	0.0	0.0	0.0	0.2	1.0	2.0	4.5
Component B current (A)	−4.0	−3.5	−3.0	−2.0	0.0	2.0	3.0	3.5	4.0

a) What type of component is component A? Give a reason for your answer.

....

....
[2]

b) Plot a current-potential difference (*I-V*) graph for component B on the grid below.
[3]

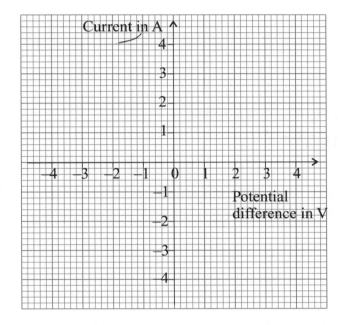

c) i) Suggest what type of component component B is. ....
[1]

ii) Explain the changes, in terms of ions and electrons, in the gradient of the *I-V* graph for component B between 0 V and 4 V.

current ↑

temperature ↑   particles vibrate move

resistance ↓

....

....
[4]

[Total 10 marks]

Score:

17

Circuit Devices

1   A student does some experiments into the properties of different circuit devices.

a) He builds the circuit shown in the diagram.

cell

thermistor

light-emitting
diode (LED)

   i) Describe and explain how the circuit current changes as the room temperature increases.

   temp ↑ = current ↑ = resistance ↓

   [2]

   ii) How would the student know that current is flowing through the LED in the forward direction?

   [1]

   iii) Give **one** use of a thermistor in a device.

   thermostat

   [1]

   iv) The cell is replaced with an alternating current (a.c.) power source. The a.c. supply undergoes *half-wave rectification*. Explain what is meant by *half-wave rectification*.

   [2]

b) The student replaces the thermistor with a light-dependent resistor (LDR) and the LED with a filament lamp.

   i) Give **one** difference between a thermistor and a light-dependent resistor.

   [1]

   ii) Give **one** use of a light-dependent resistor in an appliance.

   [1]

   iii) Give **one** disadvantage of using a filament lamp instead of an LED for lighting.

   ↑ current

   [1]

   [Total 9 marks]

Score:

9

Series Circuits

1   Twelve identical fairy light bulbs are wired in series with a 12 V battery.

a) i) Suggest **one** disadvantage of wiring the fairy lights in series.

....

....
[1]

ii) The current through one of the bulbs is 0.5 A.
Calculate the total resistance, in ohms, of the circuit.

Remember the formula connecting V, I and R.

$$V = IR$$

$$R = \frac{V}{I} = 12 \times 2$$

Resistance = ...2̶4̶.......... Ω
[2]

iii) Calculate the potential difference, in volts, across one bulb.

Potential difference = ...1̶.......... V
[1]

b) The number of bulbs in the circuit is reduced to ten.
Calculate the current, in amps, through one of the bulbs.

The total resistance of the circuit was ....2̶4̶.......... for 12 bulbs.

The bulbs are identical, so the resistance of one bulb is ....2̶.......... .

So the total resistance of the new circuit is 10 × ....2̶.......... = ...2̶0̶..........

$$V = IR$$

$$24$$

$$20$$

[3]
[Total 7 marks]

2 A student builds the circuit shown in the diagram.

series
current
same

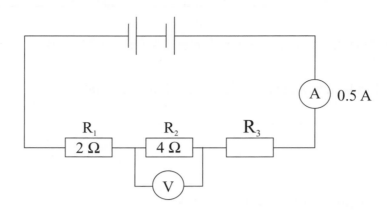

a) The total potential difference across the two cells is 12 V.

What potential difference ratings could the cells have?

Draw rings around **two** answers.

      3 V   4 V   5 V   7 V

[1]

b) The total resistance of the circuit is 24 Ω. Calculate the resistance of resistor R_3.

$$V = IR$$

$$\frac{V}{R} = 1 \qquad \frac{12}{24} = 0.5$$

Resistance of R_3 = ... Ω

[2]

c) i) What would you expect the reading on the voltmeter, in volts, to be?

$$V = IR \;\cancel{+}\; \cancel{8} \times 0.5$$

Voltmeter reading = ... V

[2]

ii) Explain why the voltmeter should not be connected in series
with the component it is measuring.

~~Volt~~ In series, voltage adds up
spread
across
all
component

[1]

[Total 6 marks]

Pd - Parralel - same
Series adds up

Score: ☐

13

Section Four — Electricity

Parallel Circuits

1 Many circuits in real life are wired in parallel.

a) A teacher asks her students to give her some real-life examples of parallel circuits.

My set of USB lights all turn on at once. If one of the bulbs breaks, all of the bulbs stop working.

The windscreen wipers, headlights and air conditioning can all be turned on and off separately in my dad's car.

L   **M**

Which student, **L** or **M**, is correctly describing a parallel circuit?

Write your answer in the box.

Explain your answer.

....

....

[1]

b) Student M looks at a simplified circuit diagram of a portable hair dryer, as shown.

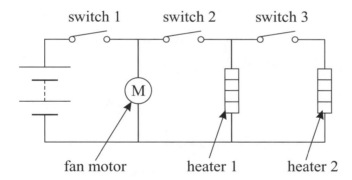

fan motor   heater 1   heater 2

The battery provides a potential difference of 12 V.

i) Complete the table by writing the potential difference across heater 1 for each combination of switch positions.

Switch 1	Switch 2	Switch 3	Potential difference across heater 1
open	open	open	0 V
closed	open	open	....
closed	closed	open	....
closed	closed	closed	....

[3]

ii) With all three switches closed in the hair dryer circuit, the current through each heater is 3 A and the current through the motor is 4 A.

Calculate the total current in the circuit, in amps.

Total circuit current = ... A

[2]

iii) Explain why ammeters must always be connected in series with the component they are measuring the current through.

....

....

....

....

[2]

c) The student builds the simple fan heater shown in the diagram.

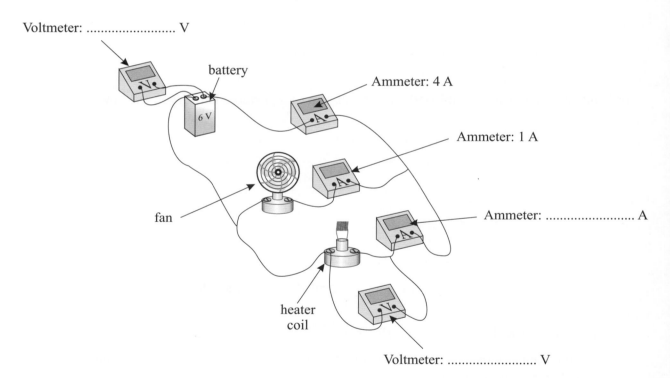

Complete the diagram by filling in the values shown on the voltmeters and ammeter.

[3]

[Total 11 marks]

Score:

11

Mains Electricity

1   The diagram shows three traces on the same cathode ray oscilloscope. The settings are the same in each case.

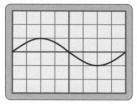      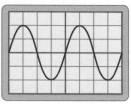

       Trace A   Trace B   Trace C

a) Which trace, A, B or C, shows the highest peak potential difference?

Write your answer in the box.

[1]

b) Which trace could represent an electricity supply produced by a battery? Explain your answer.

....

....

[2]

c) The time base of the oscilloscope is set to 5 ms per division.

   i) Calculate the frequency, in hertz, of trace C.

           5ms = 0.005s

                                     Frequency = ... Hz
                                                     [3]

  ii) Trace C is the UK mains *alternating current* (a.c.) supply.

      What is meant by *alternating current*?

      ....

                                                             [1]

  iii) The settings of the oscilloscope are altered, so the time base is 10 ms per division. The number of volts per division remains unchanged. Add to the diagram to show the trace of the UK mains supply on this altered oscilloscope.

                                       [1]

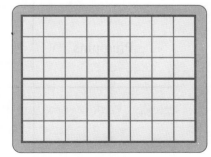

  iv) State the voltage and frequency of the UK mains electricity supply.

      ....

                         [2]

[Total 10 marks]

Score:

10

Electricity in the Home

1   Appliances must be connected to the mains electricity supply with the correct plugs and cables.

a)   A toaster is connected to the mains using a correctly wired three-pin plug, as shown in the diagram.

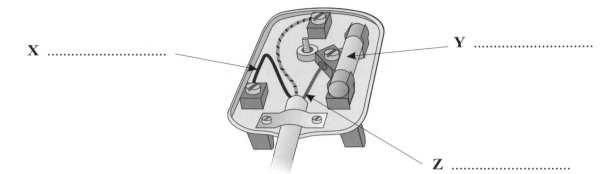

X ....

Y ....

Z ....

i)   Use words from the box to label parts **X**, **Y** and **Z** on the diagram.

live wire	neutral wire	earth wire
fuse	cable grip	brass pin

[3]

ii)   What colour(s) should the casing of the earth wire be?

....

[1]

iii) Describe the structure of the cable used to connect the toaster to the mains.

....

[1]

b)   Complete the table below with information on the materials used for different parts of the plug.

Plug part	Material used	Reason
Plug casing	Rubber or plastic	The plug casing needs to be an insulator.
Plug pins	•	....
....	Rubber or plastic	....

[2]

[Total 7 marks]

Score: ☐

7

 ☐   ☐   ☐

Fuses and Earthing

1 A student examines the wiring of a microwave oven and a vacuum cleaner. The microwave has a metal case and is wired with an earth wire for safety. The vacuum cleaner has a plastic case and is not wired with an earth wire.

a) Explain why the vacuum cleaner is still safe to use.

....

....

[2]

b) i) An electrical fault develops in which the live wire comes into contact with the metal casing of the microwave oven. Explain why this can be dangerous.

....

....

[1]

ii) The microwave oven's plug contains a fuse. Describe how the earth wire and the fuse work together to make the microwave oven safe again when the fault in part b) i) occurs.

....

....

....

[3]

iii) A washing machine has thicker cables than a microwave oven because it needs a larger power supply. Explain how this affects the current rating of the fuse that it should be fitted with.

....

[1]

c) i) Some circuits are protected by Residual Current Circuit Breakers (RCCBs) instead of a fuse and an earth wire. Explain how an RCCB protects a user when they touch a live part.

....

....

....

[3]

ii) Give **one** advantage of using an RCCB instead of a fuse and an earth wire to protect a circuit.

....

....

[1]

[Total 11 marks]

Score:

11

Energy and Power in Circuits

1 The heating element in a kettle usually contains a coil of wire made of Nichrome. The filament of a filament lamp is usually made up of a coil of tungsten wire.

a) i) Explain why the coil of wire in the heating element is designed to have a high resistance.

....

....

[1]

ii) Use a word from the box to complete the sentence.

light	heat	sound

A light bulb transfers electrical energy to useful .... energy.

[1]

iii) What energy transfer produces the most waste energy in a light bulb?

....

[1]

b) The table shows the *power* ratings and the energy transferred in one minute for three kettles.

	Kettle A	Kettle B	Kettle C
Power (kW)	2.8	....	3.0
Energy transferred in one minute (kJ)	168	150	....

i) What is meant by *power*?

....

[1]

ii) Complete the table by calculating and filling in the missing values.

[2]

iii) State which kettle will transfer the most energy in one minute. ....

[1]

iv) A student is deciding whether to buy kettle A or kettle B. She wants to buy the kettle that boils water faster. Both kettles transfer 90% of the electrical energy supplied to the water. Suggest which kettle she should choose. Give a reason for your answer.

....

....

[2]

[Total 9 marks]

Score: ☐

9

Power and Energy Change

1   A student is drilling holes and putting up some wooden shelves.
His electric drill is attached to a 12 V battery and uses a current of 2.3 A.

a) Calculate the power, in watts, of the electric drill.

Power = ... W

[2]

b) An electrical fault causes the fuse in the electric drill to blow. The student has the choice of
replacing the blown fuse with a 1 A fuse or a 5 A fuse. He decides to use the 1 A fuse as it is
rated closest to the operating current of the drill. Is he correct? Explain your answer.

....

....

....

[2]

c) When the student drills one hole, 69 C of charge passes through the motor of the drill.
Calculate the energy transferred, in joules, when drilling the hole.

Energy transferred = ... J

[2]

[Total 6 marks]

2   A student uses an electric sander to prepare wooden surfaces for painting.
The electric sander is rated at 230 V, 184 W.

a) Calculate the current, in amps, that the sander draws from the mains supply.

Current = ... A

[2]

b) Which fuse is the most appropriate for use in the electric sander?
Draw a ring around the correct answer.

       1 A   3 A   5 A   13 A   *[1]*

[Total 3 marks]

Score:

9

The Cost of Electricity

1   A student wants to work out how much energy his household could save if they turned all of their electrical appliances off overnight, instead of leaving them on standby. On Tuesday night, he leaves all the appliances on standby. He takes a meter reading at 9pm at night and at 9am the next morning. On Thursday night he repeats the experiment, but this time he turns off all the appliances.

Tuesday: Reading at 9pm	Wednesday: Reading at 9am		Thursday: Reading at 9pm	Friday: Reading at 9am
1 3 5 9 2 . 4 2 kWh	1 3 5 9 8 . 6 3 kWh		1 3 6 4 6 . 6 8 kWh	1 3 6 4 9 . 4 1 kWh

Appliances on standby   Appliances turned off

a)   Calculate the amount of energy, in kilowatt-hours, that the household saved by turning off all the electrical appliances overnight, rather than leaving them on standby.

Energy saved = ... kWh

[2]

b)   The student looks at the electricity bill shown to find out how much the household pays per kWh.

Date	Meter Reading (kWh)
11 06 13	34259
10 09 13	34783
	Cost @ 9.7p per kWh

i)   Calculate the total cost, in pence, of the bill.

Total cost = ... p

[3]

ii) The student uses a new 2.3 kW kettle to boil 1 litre of cold water. The kettle takes 180 seconds to boil. How many kilowatt-hours of energy did this use?

$$\frac{E}{t} = P$$

Number of kWh used = ...

[2]

iii) An older kettle of the same power takes longer to boil 1 litre of cold water. Will the older kettle use more, less or the same amount of energy to boil as the new kettle? Give a reason for your answer.

....

....

[2]

[Total 9 marks]

Score: ☐

9

The National Grid

1 | The National Grid is the network that is used to transmit electricity around the country.

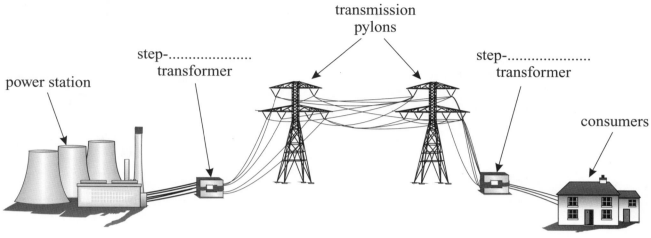

a) Complete the diagram labelling by adding **one** word to each of the two dotted lines.

[2]

b) i) Describe how energy is wasted in the National Grid.

....

....

[1]

ii) Explain the role of transformers in altering the electricity supply so that it is transmitted more efficiently in the National Grid.

....

....

....

....

[3]

c) Explain the role of transformers in supplying electricity to the consumer safely in the National Grid.

....

....

....

[2]

[Total 8 marks]

Score: ☐

8

 ☐   ☐   ☐

Magnets and Magnetic Fields

1   A student draws the magnetic field lines between four bar magnets, as shown in the diagram.

a) *In this question you will be assessed on the quality of your English, the organisation of your ideas and your use of appropriate specialist vocabulary.*

Describe an experiment that the student could have done to draw this magnetic field pattern.

magnetic field lines

magnets

....

....

....

....

....

....

....

....

....

[6]

b) Add **four** arrows to magnetic field lines on the diagram above, one between each set of magnets, to show the direction of the magnetic field.

[2]

c) The student arranges two of the magnets as shown below.

| N | S | | N | S |

i) Describe the magnetic field in the shaded region.

....

[1]

ii) State whether there will be a force of attraction, a force of repulsion, or no force between the two magnets. Give a reason for your answer.

....

....

[2]

[Total 11 marks]

Score:

11

Electromagnetism

1   A student investigated magnetic fields by passing a copper rod through a piece of flat card and connecting it in an electrical circuit, as shown in the diagram.

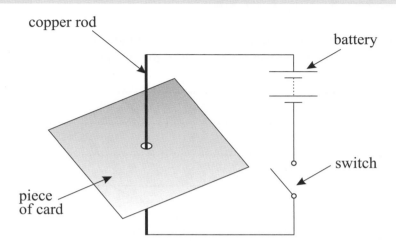

a)   Draw a ring around the correct ending to the following sentence.

When the switch is closed

> a magnetic field is created around the copper rod.
>
> a magnetic field is created by the piece of card.
>
> the magnetic field reverses direction.

[1]

b)   Some iron filings were sprinkled onto the card. When the switch was closed, a pattern developed in the iron filings. On the diagram above, sketch the magnetic field that causes this pattern. Show the direction of the magnetic field in your sketch.

[2]

c)   i)   The student removed the rod and card and attached a loop of wire passed through a piece of card to the electrical circuit, as shown below. Draw the magnetic field around the wire on the piece of card, showing the pattern and direction of the magnetic field.

[2]

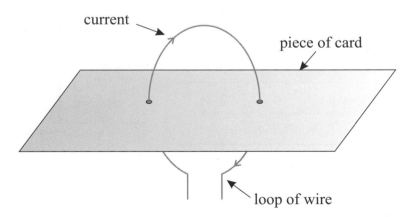

ii)   Describe how reversing the direction of the current would have affected the magnetic field.

....

[1]

[Total 6 marks]

82

2 Iron and steel are magnetic materials, which means they experience a force in a magnetic field. An electromagnet is used by a crane to lift, move and drop iron and steel.

a) Describe the basic structure of an electromagnet.

.... *coil of wire* .... *of iron core* ....

[1]

b) When a current is passed through the electromagnet, an iron bar on the ground nearby is attracted to it. When the current is stopped, the bar drops back to the ground. Explain why this happens.

....

....

....

....

[3]

c) The crane's electromagnet contains an iron core that is magnetically soft. This means that it magnetises and demagnetises very quickly.

Explain why putting a core that isn't magnetically soft in the electromagnet would cause the electromagnet to not work properly.

....

....

[1]

[Total 5 marks]

3 A simple electric bell uses an electromagnet to operate, as shown in the diagram. The spring pulls on the iron arm so that the contacts are held together when the bell is not in operation.

Explain how closing the switch causes the striker to strike the bell repeatedly, causing it to ring.

....

....

....

....

....

....

....

....

....

bell striker

iron arm

contacts

bell

electromagnet

switch

spring

cell

[Total 5 marks]

Score: ☐

16

The Motor Effect

1 The diagram shows an experiment in which a current is passed through a wire in a magnetic field.

a) When current flows through the wire, a force acts on the loop of wire and causes it to move.

   i) Explain why this force acts on the wire.

     ....

     ....

     .... *[1]*

loop of wire

current

N   S

Use Fleming's left-hand rule.

magnet

   ii) State the direction in which the loop will move.

     ....

     [1]

b) What effect will increasing the current have on the force on the loop of wire?

   ....

   [1]

c) The diagram below shows a free-rolling conducting bar on a set of fixed conducting bars in a magnetic field. All of the conducting bars have a current flowing through them. Explain why the free-rolling conducting bar doesn't move.

   ....

   ....

   ....

   ....

   [2]

conducting bar that is free to roll

magnet

magnet

N

S

current

conducting bars fixed in place

power source

[Total 5 marks]

2 The diagram shows the parts inside an earphone. Sound waves are caused by mechanical vibrations. Explain how the earphone uses an a.c. supply to produce sound waves.

coil of wire

cone

permanent magnet

base of the cone

to a.c. supply

....

....

....

....

....

....

....

[Total 4 marks]

Score:

9

The Simple Electric Motor

1   A student built a simple d.c. motor. He started by putting a loop of current-carrying wire that is free to rotate about an axis in a magnetic field, as shown in the diagram.

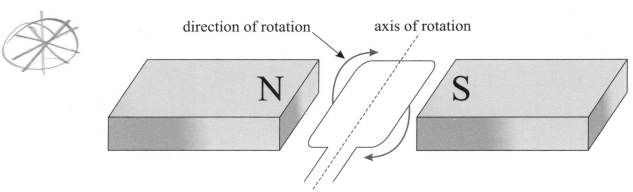

direction of rotation   axis of rotation

a) Draw an arrow on the diagram to show the direction of the current in the wire.

[1]

b) The starting position of the motor is shown in the diagram. Explain why the motor will stop rotating in the same direction after 90° of rotation from its start position.

....

....

[1]

c) Suggest and explain how the student could get the motor to keep rotating in the same direction.

....

....

[2]

d) i) Give **one** way the motor could be made to rotate faster.

....

[1]

ii) Give **two** ways in which the direction of rotation of the motor could be reversed.

1. ....

....

2. ....

....

[2]

[Total 7 marks]

Score: ☐

7

The Generator Effect

1   A student uses the rotation of a hamster wheel to power a battery charger.

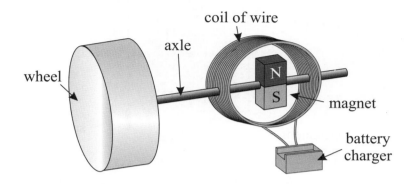

a)   Explain how rotating the wheel induces a potential difference across the battery charger.

....

....

....

[2]

b)   Describe **three** ways the potential difference induced across the battery charger could be increased.

1. ....

2. ....

3. ....

[3]

c)   Explain why the induced current in the battery charger circuit is alternating current (a.c.).

....

....

....

[2]

d)   The student removes the battery charger and leaves the ends of the wire disconnected.

Draw a ring around the correct ending to the sentence.

<table>
<tr><td rowspan="3">When the wheel rotates</td><td>there will be a potential difference induced across the ends of the wire.</td></tr>
<tr><td>a current will be induced in the wire.</td></tr>
<tr><td>no current will be induced in the wire and there will be no potential difference induced across the ends of the wire.</td></tr>
</table>

[1]

[Total 8 marks]

2   The diagram shows the internal structure of a microphone. Sound waves with a constant pitch and volume reach the microphone and cause the diaphragm to move.

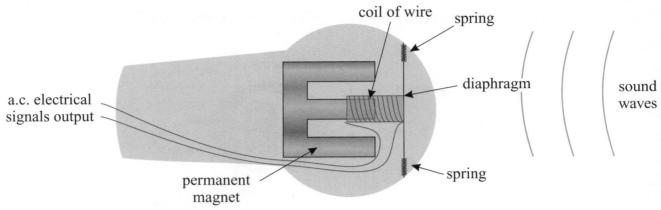

a) Explain how the microphone converts sound waves into electrical signals.

....

....

....

....

....

....

....

[3]

b) i) The pitch of the sound waves reaching the microphone is increased, meaning that there are more vibrations per second. The volume remains the same.

Explain why a higher-pitched sound would result in a higher frequency a.c. signal to be produced.

....

....

....

[2]

ii) What effect would a higher-pitched sound have on the peak potential difference of the a.c. signal produced?

....

....

[1]

[Total 6 marks]

Exam Practice Tip

The generator effect is basically the opposite of the motor effect — movement causes a potential difference and (when there is a complete circuit) a current. It's got lots of uses, usually involving turning movement into electricity, so if you're asked about a device that does that in the exam, think 'generator effect'.

Score

14

Section Five — Motors, Generators and Transformers

Generators

1 Diagram 1 shows a wind-up a.c. generator that uses the generator effect to induce an alternating current (a.c.). The generator is connected to an oscilloscope. The slip rings make sure that each end of the coil remains connected to the same oscilloscope wire. When the handle is rotated clockwise at a constant speed, the oscilloscope produces the trace shown on the screen.

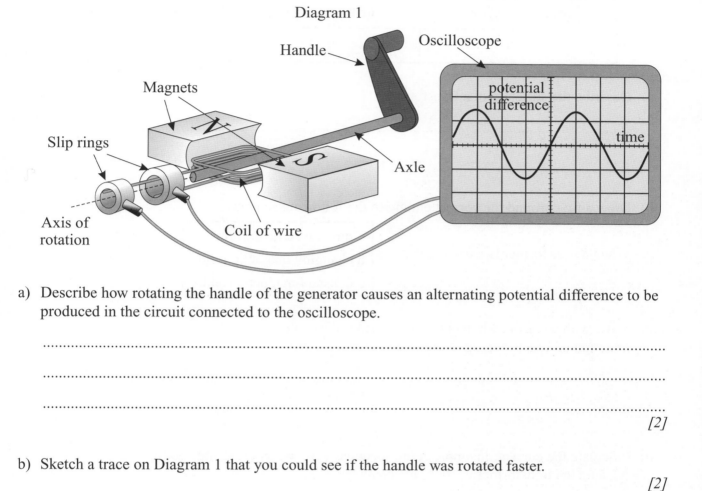

Diagram 1

a) Describe how rotating the handle of the generator causes an alternating potential difference to be produced in the circuit connected to the oscilloscope.

....

....

....

[2]

b) Sketch a trace on Diagram 1 that you could see if the handle was rotated faster.

[2]

c) Describe the trace that you would expect to see if the position of the coil was fixed and the magnets were rotated, around the axis of rotation shown in the diagram, at the original speed.

....

[1]

Diagram 2

d) Explain how the wind-up generator could be altered so that it produces the trace shown in Diagram 2.

....

....

....

....

[2]

[Total 7 marks]

Score:

Transformers

1   A student investigated a transformer by powering a spotlight
through it and measuring the quantities shown in the table.

Potential difference across primary coil (V)	Current in primary coil (A)	Potential difference across secondary coil (V)
240	0.25	12

a) Describe the structure of a transformer.

....

....
                                                                                                                            [2]

b) Draw a ring around the correct ending to the following sentence.

When a transformer is working, there is | a direct electric current
a changing magnetic field
a constant magnetic field | in the core.
                                                                                                                            [1]

c) State, with a reason, whether the transformer the student
investigated was a step-up or step-down transformer.

....

....
                                                                                                                            [1]

d) Calculate the current, in amps, in the secondary coil when using the spotlight.
Assume the transformer is 100% efficient.

Current = ... A
                                                                                                                            [2]

e) The student discovers that his laptop uses a switch mode transformer.

Describe the differences between the laptop transformer and the spotlight transformer.

....

....

....
                                                                                                                            [2]
                                                                                                                   [Total 8 marks]

2 The diagram shows a step-up transformer with 16 times more turns on its secondary coil than on its primary coil. The primary coil is connected to an alternating current (a.c.) power supply.

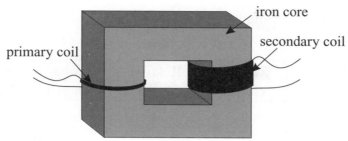

a) Explain how the transformer induces an alternating current in the secondary coil.

....

....

....

....

....

....

[4]

b) i) The potential difference across the primary coil is 25 000 V.
Calculate the potential difference, in volts, across the secondary coil.

Potential difference = ... V

[3]

ii) The output power of the transformer is 4000 kW. Calculate the input current, in amps. Assume that the transformer is 100% efficient.

Current = ... A

[2]

[Total 9 marks]

Score: ☐

17

Section Six — Nuclear Physics
Atomic Structure

1   Iodine-131 ($^{131}_{53}$I) is an unstable isotope of iodine.

a) i) Complete the table.

Particle	Relative charge	Relative mass	Number present in an atom of iodine-131
Proton	1	1	53
Neutron	0	1	53
Electron	−1	very small	131−53

[4]

ii) Name the particle(s) found in the nucleus of an atom.

....

[1]

iii) Name the particle(s) found surrounding the nucleus of an atom.

....

[1]

b) What is meant by the term *isotopes*? Place a tick in the appropriate box to indicate your answer.

☐ Atoms with the same atomic number but a different mass number.

☐ Atoms with the same mass number but a different atomic number.

☐ Atoms with the same proton number but a different atomic number.

☐ Atoms with the same number of neutrons but a different number of electrons.

[1]

c) A particle collides with an atom of iodine-131, causing it to lose an electron.

i) What was the overall charge on the iodine-131 atom before the collision?

....

[1]

ii) Name the type of particle that the atom of iodine-131 becomes as a result of the collision.

....

[1]

iii) What effect does the collision have on the atomic number and the mass number of the iodine-131? Explain your answer.

....

....

[2]

[Total 11 marks]

Score: ☐

11

Radiation

1   Uranium-235 is a radioactive isotope of uranium. It gives out radiation from its nucleus.

a) Draw a ring around the correct ending to the following sentence.

A sample of uranium-235 gives out radiation

| at random. |
| every 235 seconds. |
| only when heated in a nuclear reactor. |

[1]

b) Underground uranium-235 contributes to the low level of radiation that is present all around us all the time.

i) Give the name of this low level of radiation.

....

[1]

ii) Give **two** other sources of this low level of radiation.

1. ....

2. ....

[2]

[Total 4 marks]

2   When an unstable nucleus decays, it can give out different types of radiation.

a) Name **three** types of radiation that can be given out when unstable nuclei decay.

....

[1]

b) *In this question you will be assessed on the quality of your English,
the organisation of your ideas and your use of appropriate specialist vocabulary.*

Describe how exposure to radiation given out by unstable nuclei can be dangerous to humans.
In your description, you should write about the relative dangers of each type of radiation.

....

....

....

....

....

....

....

....

[6]

[Total 7 marks]

Score: [   ]

11

Section Six — Nuclear Physics

Ionising Radiation

1 Alpha, beta and gamma radiation are all types of nuclear radiation.
All three types of nuclear radiation can cause ionisation.

a) State which type of nuclear radiation is the most strongly ionising. Explain your answer.

....

....

[2]

b) i) Name the type of nuclear radiation that is a type of electromagnetic radiation.

....

[1]

ii) Name the type of nuclear radiation whose particles are electrons.

....

[1]

[Total 4 marks]

2 Nuclear equations can be used to show the changes in the
atomic and mass numbers of particles during alpha decay.

a) Describe the structure of an alpha particle.

....

....

[2]

b) Describe what happens to the atomic number and the mass number of a nucleus
when it undergoes alpha decay.

....

....

[2]

c) Complete this nuclear equation, which shows a polonium isotope decaying by alpha emission.

An alpha particle
can be written as
He or α.

$$^{206}_{}\text{Po} \longrightarrow \, ^{202}_{82}\text{Pb} + \, ^{\ldots}_{2}\alpha$$

[2]

[Total 6 marks]

3 A source of radiation produces two types of ionising radiation in roughly equal proportions. When a sheet of paper or a thin sheet of aluminium is placed between the source of radiation and a Geiger-Müller tube, the count rate drops to roughly half its original amount.

Suggest which **two** types of radiation are emitted by the radioactive source.

1. ....

2. ....

Give reasons for your answer.

....

....

....

[Total 3 marks]

4 Alpha particles, beta particles and gamma rays are each fired separately into a magnetic field. Their paths are shown in the diagram.

A ....   B ....   C ....

a) i) Complete the diagram by filling in the labels for each type of radiation.

[3]

ii) Which type of radiation, A, B or C, will travel the furthest in air?

....

[1]

b) Name another type of field that would cause radiation B to alter its path.

....

[1]

c) Identify the differences between paths B and C. Explain the cause of each difference.

....

....

....

....

[4]

[Total 9 marks]

Score:

22

Section Six — Nuclear Physics

94

Half-Life

1 A student measured the count rate of a radioactive sample over a period of 80 minutes. Before getting the sample out of storage she measured the count rate in the laboratory in counts per second (cps). When processing the data she subtracted this value from all her count rate readings.

a) Suggest why the student recorded the activity in the laboratory before starting the experiment.

....

....

[2]

b) The graph shows the student's processed data.

Use the graph to find the half-life of the sample, in minutes

Half-life = ... minutes

[2]

c) Use your answer to part b) to predict what the count rate of the sample will be after 120 minutes.

Count rate = ... counts per second

[2]

d) All of the radioactive samples that the laboratory uses were identical when new. The student decides to repeat the experiment with a sample that is much older than the first. Explain what effect this will have on the data collected.

....

....

....

[3]

[Total 9 marks]

2 Use words from the box to complete the sentences.

| electrons | decrease | double | stay the same |
| nuclei | halve | increase | protons | fall to zero |

The radioactivity of a radioactive isotope sample will always .... over time. The half-life is the average time taken for the number of .... of the radioactive isotope in the sample to ....

[Total 3 marks]

3 A teacher is conducting an experiment with a radioactive isotope sample. The sample has a half-life of 40 minutes.

a) i) The initial count rate of the sample is 8000 counts per second. Calculate the count rate, in counts per second, after 2 hours.

Count rate = ... cps
[2]

ii) Calculate the number of whole hours it would take for the count rate to fall below 200 cps.

Time = ... hours
[2]

b) The teacher brings out another sample of the same radioactive isotope. She tells the students that this second sample contains half as many radioactive nuclei as the first. She asks what they can tell her about its current count rate in comparison to that of the first sample. The students make the following statements.

The count rate will be the same because it's a sample of the same radioactive isotope.

A

The count rate will be lower because it contains less of the radioactive isotope.

B

The count rate will be higher because it contains more decayed nuclei.

C

Which student, A, B, or C, made a **correct** statement?

....
[1]
[Total 5 marks]

Exam Practice Tip

There's quite a lot of maths involved in half-life questions. Try not to panic though. Just take your time and go through the different stages slowly. Make sure you show your workings for calculations or if you've used a graph. You then might still be able to pick up marks even if you get the final answer wrong. Bonus.

Score

17

Uses of Radiation

1   In hospitals, radioactive materials are used in a wide variety of situations.

a) Before a surgical instrument is used in an operation, it is placed in a machine that exposes it to high-intensity gamma rays. Suggest why.

....

                                                   [1]

b) The table shows six different radioactive isotopes.

Isotope	Type of radiation emitted	Half-life
Rhenium-186	beta	4 days
Technetium-99m	gamma	6 hours
Actinium-225	alpha	10 days
Polonium-206	beta	9 days
Barium-137m	gamma	2.5 minutes

i) Radioactive isotopes can be used as medical tracers. They are injected into the body or swallowed and then their progress around the body can be tracked over time using an external detector. Suggest which isotope would be the most useful for this purpose and explain why.

....

....

....

                                                   [3]

ii) Radiotherapy is used to treat cancer patients. Some radiotherapy uses cobalt-60, a radioactive isotope that emits gamma rays. Explain how radioactive isotopes such as cobalt-60 are used in the treatment of cancer.

....

....

....

                                                   [2]

iii) Explain why exposure to radioactive isotopes must be kept to a minimum during treatment.

....

....

....

                                                   [2]

[Total 8 marks]

2 In some smoke detectors, a source of alpha radiation is used to ionise the air particles between two electrodes, allowing a current to flow. If smoke particles pass between the electrodes they disrupt the current, triggering an alarm.

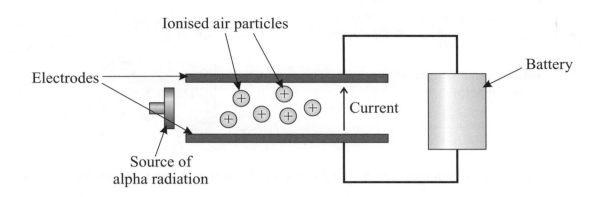

a) Suggest why the smoke detector uses alpha radiation instead of beta or gamma radiation.

....

....

....

....

[2]

b) Explain why the distance between the radioactive source and the electrodes must be made as small as possible.

....

....

[1]

c) The table shows three radioactive isotopes.

Isotope	Radiation emitted	Half-life
Polonium-211	alpha	0.516 seconds
Americium-241	alpha	433 years
Uranium-229	alpha, beta	58.3 minutes

Suggest which would be the most suitable for use in a smoke detector and explain why.

....

....

....

[3]

[Total 6 marks]

Score: [　　]

14

Nuclear Fission and Fusion

1 Nuclear fission takes place in nuclear reactors. The diagram shows the basic structure of a gas-cooled nuclear reactor.

a) Give **one** fuel that can be used in a nuclear reactor.

....
[1]

b) i) Describe what happens during a single nuclear fission event and the products formed.

....

....

....

....
[4]

ii) Sketch a diagram to show how a fission chain reaction happens.
Label one of each reactant and product in the reaction.

[3]
[Total 8 marks]

2 Stars such as the Sun release vast amounts of energy.

a) Name the process by which energy is released in stars.

....
[1]

b) Explain how all the naturally occurring elements in the Universe were formed from hydrogen by this process, and how they were distributed throughout the Universe.

....

....

....
[3]
[Total 4 marks]

Score: ☐

12

The Life Cycle of Stars

1 Stars are formed when clouds of gas and dust are pulled together.

a) i) Name the force that causes this to happen. ....

[1]

ii) Give **one** object, other than a star, that can be formed by this process.

....

[1]

b) Main sequence stars undergo a long stable period lasting millions of years.

i) Explain why stars are stable during their main sequence stage.

....

[1]

ii) Explain why main sequence stars are able to maintain their energy output for such long periods of time.

....

....

[1]

c) The life cycle of a star depends on its size.

i) Describe the stages in the life cycle of a star the size of our Sun after its main sequence stage has ended.

....

....

....

[3]

ii) The diagram shows the life cycle of a star much larger than our Sun after its main sequence stage. Complete the diagram by labelling the stages A, B and C of the life cycle of the star.

Main sequence star   A ....   Supernova   B ....

C .... [3]

iii) In which stage in the life cycle of this star are elements heavier than iron formed?

....

[1]

[Total 11 marks]

Score:

11

Candidate Surname		Candidate Forename(s)	

Centre Number	Candidate Number

Level 1/2 Certificate in Physics
Paper 1

Practice Paper
Time allowed: 90 minutes

You must have:
- A pencil.
- A calculator.
- A ruler.

Total marks:

Instructions to candidates
- Use **black** ink to write your answers.
- Write your name and other details in the spaces provided above.
- Answer **all** questions in the spaces provided.
- In calculations, show clearly how you worked out your answers.

Information for candidates
- The marks available are given in brackets at the end of each question.
- There are 90 marks available for this paper.
- You might find the equations on pages 155-156 useful.
- You should answer Question 8 (c) with continuous prose.
  You will be assessed on the quality of your English,
  the organisation of your ideas and your use of
  appropriate specialist vocabulary.

Advice for candidates
- Read all the questions carefully.
- Write your answers as clearly and neatly as possible.
- Keep in mind how much time you have left.

Get the answers

Your free Online Edition of this book includes the complete answers and mark scheme for this Exam Paper —
you can even print them out. There's more info about how to get your Online Edition at the front of this book.

Answer **all** questions

1   A student did an experiment using the apparatus shown.
He used identical electric heating coils to heat a beaker of water and a beaker of oil.
He used 1 kg of each liquid.

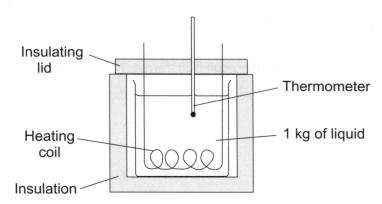

The student took the temperatures of both the liquids before heating, and then again after ten minutes of heating. His results are shown in the table.

	Water	Oil
Initial temperature (°C)	18	16
Final temperature (°C)	48	88

1 **(a)**   The student noticed that the temperatures of the liquids decreased after the heaters were switched off.

Explain why this happened.

....
[1]

1 **(b)**   Suggest how the student could have improved the validity of his data.

....

....
[1]

Question 1 continues on the next page

Turn over ▶

1 (c) (i) Which of the liquids has a lower specific heat capacity? ....

Give the reason for your answer.

....

....

[1]

1 (c) (ii) During the experiment, the heating coil transferred 126 kJ of energy to each liquid.

Use data from the experiment to calculate the specific heat capacity of oil.

....

....

....

....

....

Specific heat capacity of oil = ... kJ/kg°C

[3]

1 (d) Both oil and water can be used to carry heat around heating systems.

Most heating systems use water rather than oil. Use the results of this experiment to suggest why.

....

....

....

[2]

[Total 8 marks]

2   The diagram below shows a crane picking up a metal anvil using an electromagnet.

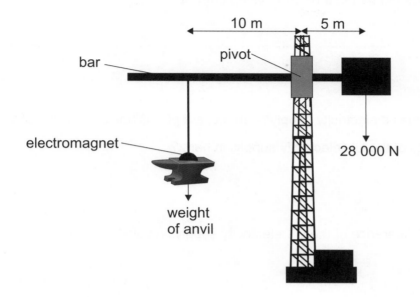

2 (a)   Use a phrase from the box to complete the sentence.

gravitational field	**magnetic field**	**free electrons**

An electromagnet is a magnet whose .... can be turned on and off.

[1]

2 (b)   The crane is balanced about the pivot. Use the information on the diagram to calculate the weight, in newtons, of the anvil. Ignore the weight of the bar in your calculation.

....

....

....

....

....

....

Weight of anvil = ... N

[4]

Question 2 continues on the next page

Turn over ▶

2 (c)   The electromagnet is powered by a generator that supplies an *alternating current* (a.c.).

What is meant by the term *alternating current*?

....

....

[1]

2 (d)   The UK mains electricity supply is an a.c. supply.   What is the value of the:

2 (d) (i)   frequency of the UK electricity supply, in hertz?

.... Hz

[1]

2 (d) (ii) potential difference of the UK electricity supply, in volts?

.... V

[1]

[Total 8 marks]

3 A lorry and a motorbike are driving around a bend in the road.
The diagram shows a view from above the road.

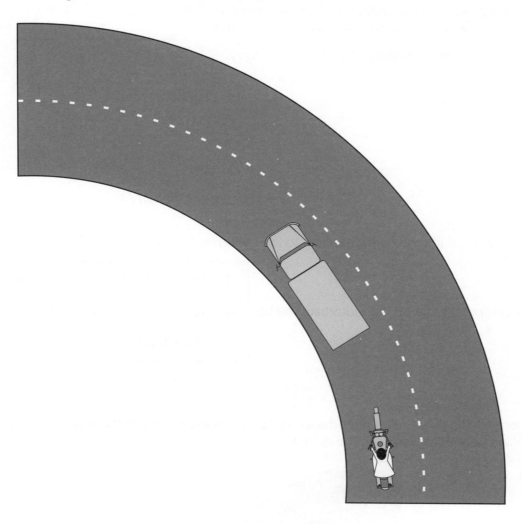

3 (a) Name the centripetal force that allows the vehicles to travel around the bend.

....
[1]

3 (b) (i) How do you know that the motorbike is accelerating, without knowing its velocity?

....

....
[1]

3 (b) (ii) Draw an arrow on the diagram to show the direction in which the motorbike is accelerating.
Label the arrow with an **A**.

[1]

Question 3 continues on the next page

Turn over ▶

3 (c)   The lorry has a mass of 8000 kg. The bike and rider have a total mass of 200 kg.
Both the lorry and motorbike are travelling at 12 m/s.

3 (c) (i) Calculate the kinetic energy, in joules, of the lorry.

....

....

....

Kinetic energy of the lorry = ... J

[2]

3 (c) (ii) Draw a ring around the correct ending to the sentence below.

The centripetal force on the motorbike will be

| the same as the force on the lorry. |
| greater than the force on the lorry. |
| less than the force on the lorry. |

[1]

3 (c) (iii) The lorry slows down. The motorbike overtakes the lorry in the outside lane without
changing speed, as shown in the diagram.

Describe how the centripetal force on the motorbike has changed.

Give a reason for your answer.

....

....

....

[2]

3 (d)   The lorry has many safety features, including air bags and seat belts.

air bag

seat belt

3 (d) (i)   Explain, in terms of momentum, how air bags help protect passengers during a crash.

....

....

....

....

[3]

3 (d) (ii)   Seat belts are made from an elastic material. They stretch slightly during a collision to reduce the risk of injury on the passengers.

The material used for one type of seat belt has a spring constant of 180 000 N/m. Calculate the extension, in metres, of the seat belt if the force on a passenger in a crash is 13 500 N.

....

....

....

....

....

Extension = ... m

[2]

[Total 13 marks]

Turn over for the next Question

Turn over ▶

4 A home owner wants to reduce her energy bills.

The table below shows the data she has collected on home insulation she is thinking of installing.

Letter	Home insulation	Cost to buy	Annual saving
A	Loft insulation	£300	£150
B	Solid wall insulation	£18 000	£900
C	Draught proofing	£300	£55

4 (a) Which type of insulation would be the most cost effective over a period of two years?

Write your answer in the box.

Explain your answer.

....

....

[3]

4 (b) The home owner also decides to replace her washing machine with a more efficient model.

4 (b) (i) Give **one** useful energy transfer that occurs in a washing machine.

....

[1]

4 (b) (ii) Describe what usually happens to the energy 'wasted' by an appliance and why it becomes less useful.

....

....

....

[2]

4 (c) The washing machine has a power rating of 540 W and an efficiency of 35%. Calculate the useful power output, in watts, of the washing machine.

....

....

....

Useful power output = ... W

[2]

[Total 8 marks]

5   A student used data from the Highway Code to produce the following graph of *stopping distance* against speed for a typical car travelling on a dry road.

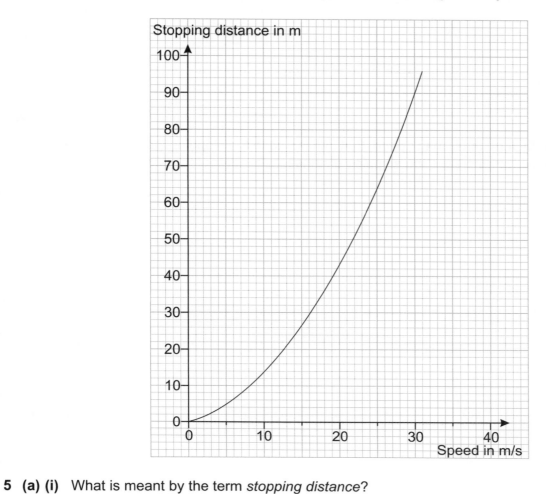

5 (a) (i)   What is meant by the term *stopping distance*?

....

....

[1]

5 (a) (ii) Use the graph to find the typical stopping distance, in metres, of a car travelling at 18 m/s on a dry road.

Stopping distance = ... m

[1]

5 (b)   The data used to plot the graph was obtained by observing a large number of drivers and vehicles. Explain why it was sensible to collect the data this way.

....

....

....

[2]

Question 5 continues on the next page

Turn over ▶

5 (c) Describe the relationship between the speed of a car and the braking force required for it to stop in a certain distance.

....

....

....

[2]

5 (d) Give **two** factors, other than speed, that can increase the stopping distance of a car and explain how they increase the stopping distance.

1. ....

....

....

2. ....

....

....

[4]

[Total 10 marks]

6   The table shows data on the radioactive decay of six radioactive isotopes.

Isotope	Symbol	Type of decay	Half-life
Radium-226	$^{226}_{88}$Ra	alpha	1600 years
Radon-222	$^{222}_{86}$Rn	alpha	3.8 days
Polonium-218	$^{218}_{84}$Po	alpha	3 minutes
Bismuth-214	$^{214}_{83}$Bi	beta	20 minutes
Lead-210	$^{210}_{82}$Pb	beta	22 years
Bismuth-210	$^{210}_{83}$Bi	alpha, beta	5 days

6 (a) (i)   What is the atomic number of bismuth-214?

....

[1]

6 (a) (ii)   How many neutrons does bismuth-214 have?

....

[1]

6 (b)   Complete the following sentence.

Isotopes of the same element have the same number of protons but a different

number of ....

[1]

Question 6 continues on the next page

Turn over ▶

6 (c) Radium-226 and lead-210 are both radioactive isotopes that decay and emit radiation.

6 (c) (i) Using data from the table, complete the equation to show a decay of radium-226.

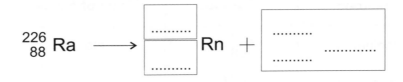

$$^{226}_{88}\text{Ra} \longrightarrow \boxed{\begin{array}{c}\ldots\ldots\ldots \\ \ldots\ldots\ldots\end{array}}\text{Rn} + \boxed{\begin{array}{c}\ldots\ldots\ldots \\ \qquad\ldots\ldots\ldots \\ \ldots\ldots\ldots\end{array}}$$

[3]

6 (c) (ii) Using data from the table, complete the equation to show a decay of lead-210.

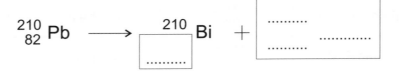

$$^{210}_{82}\text{Pb} \longrightarrow {}^{210}_{\boxed{\ldots\ldots\ldots}}\text{Bi} + \boxed{\begin{array}{c}\ldots\ldots\ldots \\ \qquad\ldots\ldots\ldots \\ \ldots\ldots\ldots\end{array}}$$

[2]

6 (d) The activity of a sample of polonium-218 is 80 counts per second.
Using data from the table, calculate how long, in minutes, it will take
for the activity of the sample to drop to 5 counts per second.

....

....

....

....

Time taken = ... minutes

[2]

6 (e) Describe **one** use of a radioactive isotope.

....

....

....

[3]
[Total 13 marks]

7 A student used a converging lens to look at an object.
The distance between the lens and the object was less than the focal length of the lens.
A diagram of the object, image and lens is shown below. The diagram is drawn to scale.

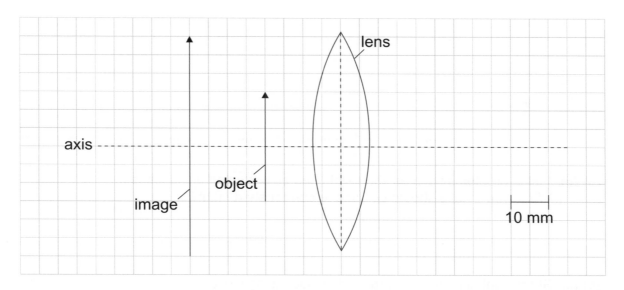

7 (a) Give **two** ways in which the nature of the image produced would differ if the object was placed at a distance greater than the focal length of the lens.

1. ....

2. ....

[2]

7 (b) (i) Use the diagram to find the focal length of the lens, in centimetres.

Focal length = ... cm

[1]

7 (b) (ii) Calculate the magnification of the lens using the heights of the object and image in the diagram above.

....

....

....

....

Magnification = ...

[2]

Question 7 continues on the next page

Turn over ▶

7 (c) Short-sighted people cannot focus on distant objects.
Short sight can be corrected by using lenses.

7 (c) (i) Describe the causes of short sight.

....

....

....

....

[2]

7 (c) (ii) Vision problems such as short sight can be corrected using lenses.
What type of lens should be used to correct short sight?

....

[1]

7 (c) (iii) Short sight can also be corrected without using lenses.
Name **one** method used to correct short sight without using a lens.

....

[1]

7 (d) The diagram below shows the basic structure of the eye.

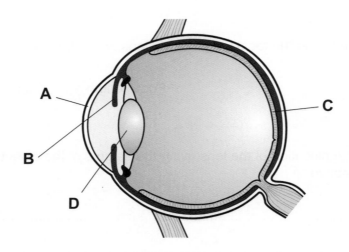

7 (d) (i) The table gives some information about the parts of the eye labelled in the diagram above.

Complete the table by filling in the missing information.

Structure	Letter	Function
Cornea	A	Focuses light from objects into the eye.
Retina	....	....
Iris	....	....

[3]

7 (d) (ii) Explain how the lens, ciliary muscles and suspensory ligaments work together to focus light from objects at varying distances.

....

....

....

....

....

[4]

[Total 16 marks]

Turn over for the next Question

Turn over ▶

8 A designer creates a metal soup bowl with a metal lid. He pours hot soup into the bowl.

8 (a) (i) What method of heat transfer allows heat to be transferred from the soup through the metal bowl?

Draw a ring around the correct answer.

conduction **convection** **radiation**

[1]

8 (a) (ii) Complete the sentence below about how heat is transferred through the metal bowl.

Metal contains .... electrons. The

heat from the soup causes these electrons to move faster, so they collide

with the metal atoms of the bowl more often. These collisions transfer

.... to the metal atoms, heating the bowl.

[2]

8 (b) (i) The designer leaves some hot soup in the metal soup bowl without a lid for an hour. Some of the soup evaporates.

What effect will the evaporation have on the temperature of the remaining soup?

....

[1]

8 (b) (ii) Explain how putting the lid on a bowl of hot soup would help to keep the soup warm for longer.

....

....

....

....

....

[4]

8 (c) *In this question you will be assessed on the quality of your English, the organisation of your ideas and your use of appropriate specialist vocabulary.*

Designers can also use metal to create products that help to keep objects cool.

The diagram shows a metal cooling fin designed to be used on a motorbike engine.

Explain how the design of the metal cooling fin helps it maximise energy transfer from the hot motorbike engine.

....

....

....

....

....

....

....

....

....

....

....

[6]

[Total 14 marks]

END OF QUESTIONS

Candidate Surname		Candidate Forename(s)

Centre Number	Candidate Number

Level 1/2 Certificate in Physics
Paper 2

Practice Paper
Time allowed: 90 minutes

You must have:
- A pencil.
- A ruler.
- A calculator.

Total marks:

Instructions to candidates
- Use **black** ink to write your answers.
- Write your name and other details in the spaces provided above.
- Answer **all** questions in the spaces provided.
- In calculations, show clearly how you worked out your answers.

Information for candidates
- The marks available are given in brackets at the end of each question.
- There are 90 marks available for this paper.
- You might find the equations on pages 155-156 useful.
- You should answer Question 3 (a) with continuous prose.
  You will be assessed on the quality of your English,
  the organisation of your ideas and your use of
  appropriate specialist vocabulary.

Advice for candidates
- Read all the questions carefully.
- Write your answers as clearly and neatly as possible.
- Keep in mind how much time you have left.

Answer **all** questions

1   The diagram shows a hot water tank used to heat and store water.

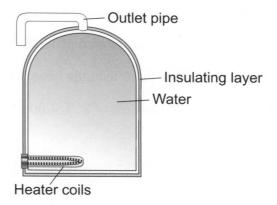

1 (a)   Describe how heat is transferred throughout the water in the tank.

....

....

....

....

....

....

[4]

1 (b)   A student wanted to investigate how the thickness of an insulating layer affects how quickly the water cools. She carried out an investigation to test how the thickness of a cotton wool jacket affects its ability to insulate. The diagram shows the apparatus she used.

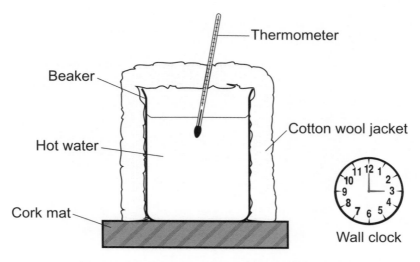

Question 1 continues on the next page

Turn over ▶

She used the following method:

- Put 200 cm³ of boiling water in the glass beaker.

- Fit a 1 cm thick cotton wool jacket over the beaker.

- Put a thermometer into the beaker through the cotton wool jacket.

- Record the time when the temperature cools to 95 °C.

- Record the temperature after three minutes have passed from this time.

- Repeat the experiment using jackets of 2, 3, 4 and 5 cm thickness.

1 (b) (i) What is the independent variable in this investigation?

....

[1]

1 (b) (ii) Circle the correct word from the box to complete the passage below.

When the student repeated the investigation she got very similar results.

It would be reasonable to say that any errors she made were probably

random
systematic
anomalous

.

[1]

1 (b) (iii) Suggest **one** way in which the precision of the student's measurements could be improved.

....

....

[1]

1 (c) The graph shows the student's results.

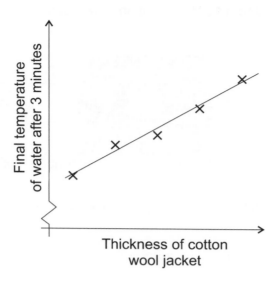

Thickness of cotton
wool jacket

Use the graph to write a suitable conclusion for this investigation.

....

....

[1]

1 (d) Explain how the arrangement and movement of particles in a material determines
whether it is an insulator or a conductor.

....

....

....

....

[3]

[Total 11 marks]

Turn over for the next Question

Turn over ▶

2   A student made a simple transformer from an iron core and two lengths of wire, as shown in the diagram. He connected a 12 V alternating power supply to one of the coils and a lamp and voltmeter to the other coil.

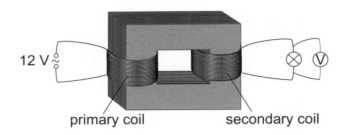

12 V

primary coil   secondary coil

2 (a)   Explain how a potential difference is generated across the secondary coil.

....

....

....

[2]

2 (b)   The student experimented with the transformer by changing the number of turns on the secondary coil and measuring the potential difference across the lamp each time. His results are shown in the table.

Number of turns on secondary coil (n_s)	5	10	15	20	25
Potential difference induced across the lamp (V_s)	3.8	7.5	1.3	15	18.8

Use the grid below to plot the student's results. Draw a line of best fit.

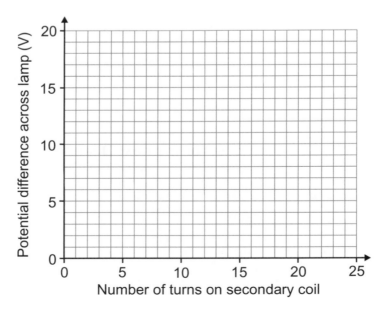

[2]

2 (c) The student thinks he may have recorded an anomalous result.

2 (c) (i) Circle the anomalous result on the graph.
Suggest **one** reason why this anomalous result could have occurred.

....

....

[2]

2 (c) (ii) Use your graph to estimate the actual potential difference induced for this number of turns on the secondary coil.

....

[1]

2 (d) Calculate the number of turns on the primary coil using data from the table.
Assume the transformer is 100% efficient.

....

....

....

Number of turns on the primary coil = ... turns

[2]

2 (e) (i) Traditional step-up and step-down transformers are used within the National Grid to alter the voltage and current of electricity. Switch mode transformers are used in mobile phone chargers.

Explain why switch mode transformers are more suited to this use than traditional transformers.

....

....

....

....

[2]

2 (e) (ii) Complete the following sentences.

Step-up transformers have .... turns on the secondary coil than

the primary coil. This results in a .... potential difference across

the secondary coil compared to the primary coil.

[2]

[Total 13 marks]

Turn over for the next Question

Turn over ▶

3   A student did an experiment to measure the velocity of a ball as it hit the ground when dropped from various heights. She used a ball, a tape measure, and a velocity sensor connected to a computer. The graph shows the data she collected.

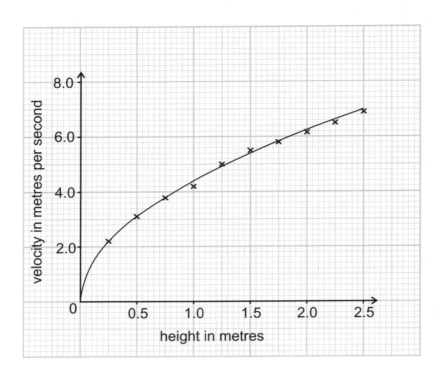

3 (a)   *In this question you will be assessed on the quality of your English, the organisation of your ideas and your use of appropriate specialist vocabulary.*

Describe how the student might have collected the data shown in the graph.
Your answer should include ways in which she could have ensured her experiment was a fair test.

....

....

....

....

....

....

....

....

....

....

[6]

3 (b) (i) Give **one** advantage of using a computer to measure and record data.

....

....

[1]

3 (b) (ii) The student measured the velocity of the ball dropped from each height five times, then used this data to calculate an average. Suggest why.

....

....

....

....

[2]

3 (c) The ball accelerated towards the ground at 9.8 m/s^2 due to gravity. The weight of the ball was 0.196 N. Calculate the ball's mass, in kilograms.

....

....

....

Mass = ... kg

[2]

[Total 11 marks]

Turn over for the next Question

Turn over ▶

4   Sunscreens absorb ultraviolet (UV) radiation
from the Sun to stop it reaching the skin.

4 (a)   Describe how ultraviolet radiation can be dangerous to human health.

....

....

....

[2]

4 (b)   The diagram shows the wavelengths of the waves that make up the electromagnetic spectrum.

wave type	A	B	C	D	E	F	G
wavelength in metres	1 to 10^4	10^{-2}	10^{-5}	10^{-7}	10^{-8}	10^{-10}	10^{-15}

Increasing energy

4 (b) (i)   Which letter corresponds to ultraviolet radiation?

....

[1]

4 (b) (ii)   Name **one** other type of electromagnetic wave.

....

[1]

4 (b) (iii)   State **one** use of ultraviolet waves.

....

[1]

4 (c)   A student carried out an investigation to test the amount of UV radiation blocked by different sunscreens. He placed a sample of each sunscreen on a separate microscope slide, then placed the slides, one at a time, between a UV source and a UV sensor. He then recorded the amount of UV radiation that was able to pass through each sample of sunscreen.

The equipment used by the student is shown in the diagram.

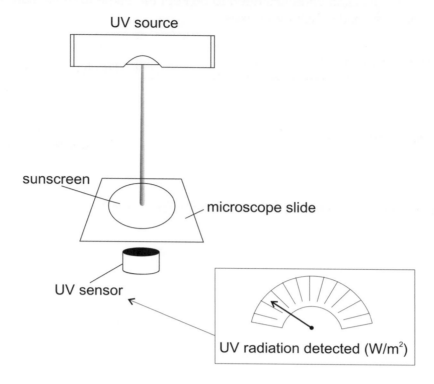

The student also tested a control microscope slide, which had no sunscreen on it. The results are shown in Table 1.

Table 1

Sunscreen sample	Reading on UV sensor in watts per metre squared			
	Reading 1	Reading 2	Reading 3	Average
control	49	50	51	50
1	13	13	10	12
2	14	13	18	....
3	18	16	17	17
4	21	22	17	20
5	23	24	22	23
6	23	23	26	24

Question 4 continues on the next page

Turn over ▶

4 (c) (i) Which of the following variables need to be kept the same to make sure the results of the experiment are valid? Tick **two** boxes.

☐ The thickness of the layer of sunscreen on the microscope slide.

☐ The type of sunscreen used.

☐ The length of the microscope slide.

☐ The distance between the UV source and the sensor.

[1]

4 (c) (ii) Why is it necessary to set up a control microscope slide?

....

....

....

....

[2]

4 (c) (iii) Complete Table 1 by calculating an average UV sensor reading for sunscreen sample 2.

....

....

[1]

4 (d) Which sunscreen sample tested provides the best protection from UV radiation? Explain your answer.

....

....

....

[2]

4 (e) Table 2 shows the concentration of titanium dioxide in the 6 sunscreens.

Table 2

Sunscreen sample	1	2	3	4	5	6
Concentration of titanium dioxide (% composition by mass)	25	18	16	13	10	9

4 (e) (i) What can you conclude from the data in Table 1 and Table 2?

....

....

[1]

4 (e) (ii) Suggest **one** way in which the validity of this conclusion could be improved.

....

....

[1]

4 (f) Student A normally gets sunburn after 30 minutes in the direct sunlight on a hot day.
Student B normally gets sunburn after 60 minutes in the direct sunlight on a hot day.

Suggest why student A might wear a sunscreen with a higher concentration
of titanium dioxide than student B.

....

....

[1]

[Total 14 marks]

Turn over for the next Question

Turn over ▶

5   A student investigated what affected the size of the potential difference induced in a coil of wire when rotating it in a magnetic field. He used a variable-speed motor to rotate different coils of wire between the poles of a horseshoe magnet, and recorded the peak potential difference induced across a resistor attached to the coil.
The equipment he used is shown in Diagram 1.

Diagram 1

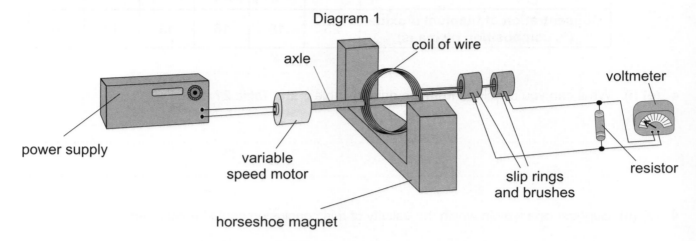

The student compared the peak potential difference induced using two different coils, A and B, shown in Diagram 2. The coils were made from the same wire and rotated at the same speed, in the same position and inside the same horseshoe magnet.
The student's results are shown in Graph 1.

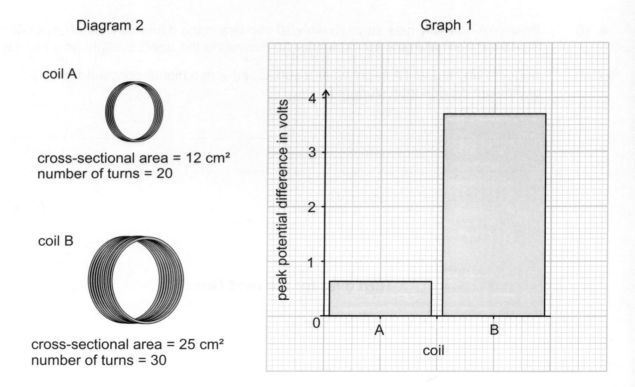

Diagram 2

coil A

cross-sectional area = 12 cm²
number of turns = 20

coil B

cross-sectional area = 25 cm²
number of turns = 30

Graph 1

5 (a)   The student concluded from these results that, for all coils of wire, the more turns in the coil, the higher the peak potential difference induced. Explain why his conclusion is not valid.

....

....

....

[2]

5 (b) (i) The student also investigated the effect of changing the speed of rotation on the peak potential difference induced in a coil of wire. He tested three different coils, C, D and E, made of the same wire in the same horseshoe magnet at five different motor speeds. The number of turns and the cross-sectional areas of the three coils are given in Table 1. His results are shown in Graph 2.

Table 1

	Number of turns	Cross-sectional area (cm²)
Coil C	10	12
Coil D	15	12
Coil E	20	12

Graph 2

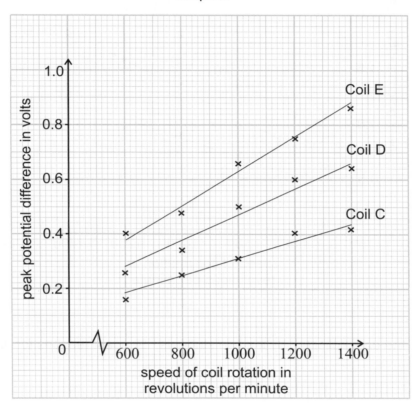

Describe the student's results in Graph 2.

....

....

....

....

[2]

Question 5 continues on the next page

Turn over ▶

5 (b) (ii) Another student carried out a similar experiment using the same coils of wire C, D and E. He rotated the coils C, D and E in a magnetic field at the same speeds as the first student. The second student's results are shown in Graph 3.

Graph 3

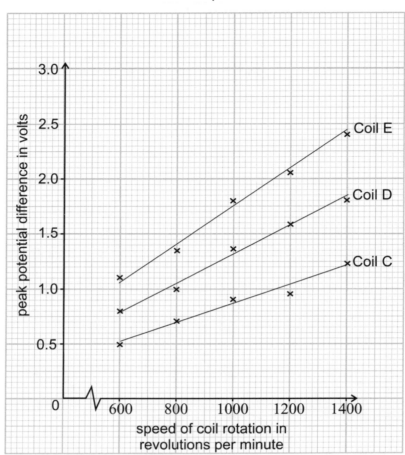

Compare the data collected by the two students. Suggest reasons for any differences.

....

....

....

....

....

[3]

[Total 7 marks]

6 A student carried out an experiment to find out what types of radiation are emitted by a source. The diagram shows three tests done by the student.

In Test 1, he placed the source in front of a detector to find the radiation, in counts per minute, emitted by the source. In Test 2, he placed a piece of paper between the source and detector. In Test 3, he removed the source to measure background radiation.

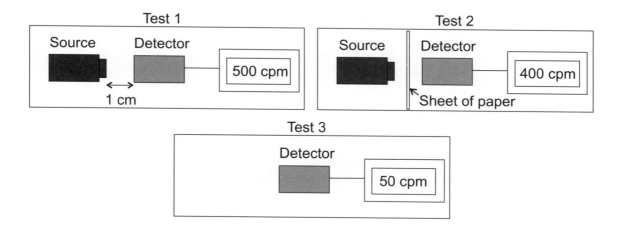

The student's results show that the source emitted more than one type of radiation.

6 (a) Name **one** type of radiation emitted by the source.
Give a reason for your answer.

....

....

....

[2]

The student pointed the source into a uniform electric field. The paths of two types of nuclear radiation, 1 and 2, are shown in the diagram.

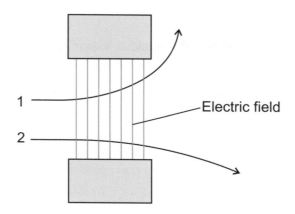

Question 6 continues on the next page

Turn over ▶

6 (b) Name the **two** types of radiation detected in this experiment. Explain why the two types of radiation are deflected in opposite directions and by different amounts as they travel through the electric field.

....

....

....

....

....

[4]

6 (c) Radioactive sources are used in nuclear power plants to help generate electricity. The diagram shows the outline of a nuclear fission reaction in a power station reactor.

Using the diagram, describe the nuclear fission reaction in a power station reactor.

....

....

....

....

....

[4]

[Total 10 marks]

7   Student A investigated how the current flowing through a filament lamp changes with the
        potential difference across it. He used the circuit shown in the diagram.

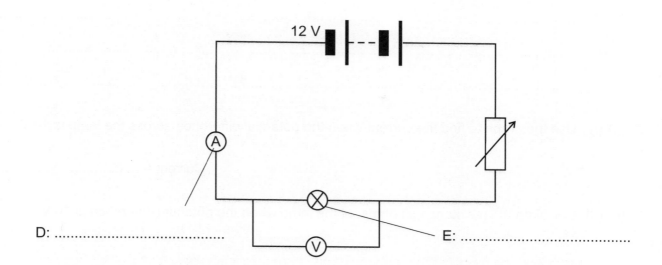

7 (a)   Label components D and E in the diagram by adding a word or phrase to each dotted line.

                                                                                        [2]

7 (b)   By adjusting the variable resistor, the student varied the potential difference across the
        lamp, and measured the current through the circuit. Table 1 and Graph 1 show his results.

Table 1

Potential difference in volts	0.0	3.0	5.0	7.0	9.0	11.0
Current in amps	0.0	1.0	1.3	1.8	1.9	2.1

Graph 1

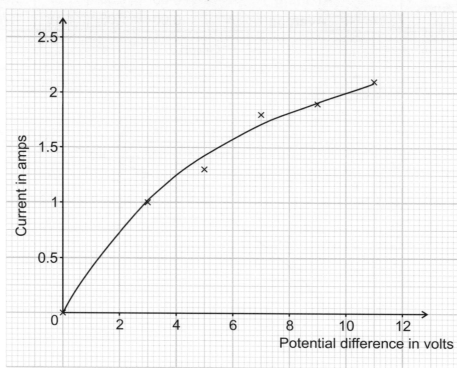

Question 7 continues on the next page

Turn over ▶

Describe and explain, in terms of particle collisions, what happens to the resistance of the lamp as the current through it increases.

....

....

....

[3]

7 (c) (i) Use the graph to find the current when the potential difference across the lamp is 10 V.

Current = ... A

[1]

7 (c) (ii) Calculate the resistance, in ohms, of the lamp when the potential difference is 10 V.

....

....

Resistance = ... Ω

[2]

7 (d) Student B conducts the same experiment. Table 2 and Graph 2 show the data he collected.

Table 2

	Potential difference in volts	0.00	3.00	5.00	7.00	9.00	11.00
Current in amps	Reading 1	0.00	0.78	1.33	1.82	1.86	2.12
	Reading 2	0.00	0.89	1.45	1.64	1.93	1.94
	Reading 3	0.00	1.33	1.57	1.64	1.91	2.09
	Average	0.00	1.00	1.45	1.70	1.90	2.05

Graph 2

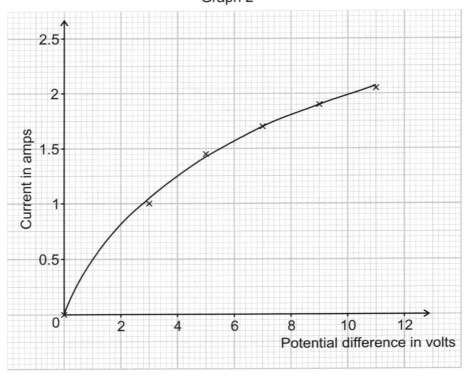

Compare and evaluate the data collected by the two students.

....

....

....

....

....

[3]

7 (e) (i) Sketch the graph you would expect to see if the lamp was replaced with a diode.

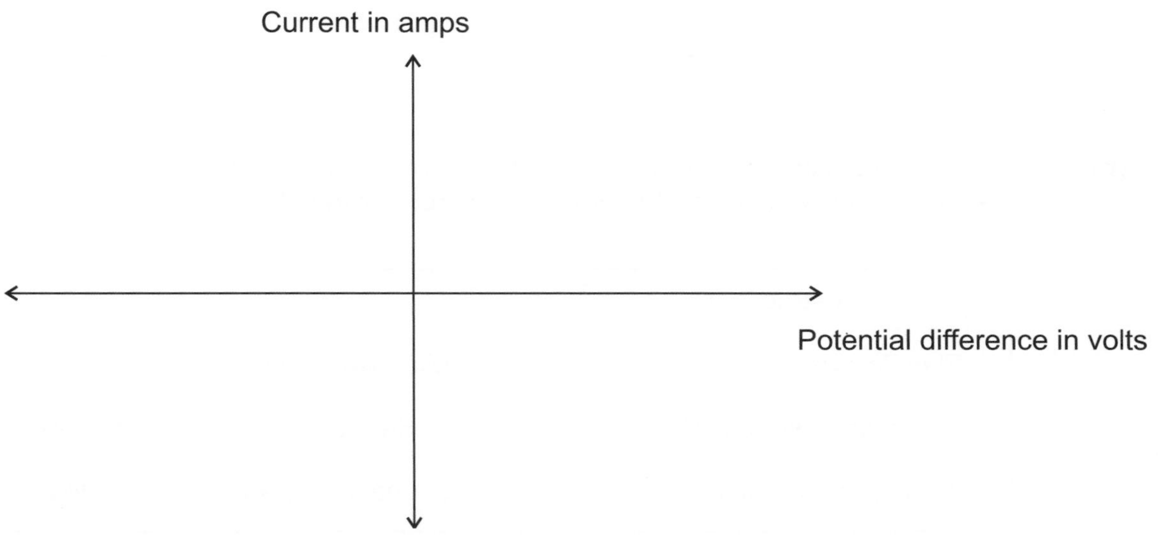

Current in amps

Potential difference in volts

[1]

7 (e) (ii) Tick the box next to the circuit symbol for a diode.

☐ ⊏▭⊐   ☐ ▱̸

☐ ─∘⁄∘─   ☐ ─⊕─

[1]

[Total 13 marks]

Turn over for the next Question

Turn over ▶

8 During the manufacturing process, a ready meal is exposed to gamma radiation.

8 (a) (i) Complete the sentence, using a word from the box.

infrared	microwave	ionising	ultraviolet

Gamma radiation is a type of .... radiation.

[1]

8 (a) (ii) Explain why ready meals are exposed to gamma radiation during manufacturing.

....

....

[1]

8 (b) The ready meal can be cooked in a microwave oven or an electric oven.
The cooking instructions for the ready meal are shown in the diagram.

Cooking instructions

Electric oven   Microwave oven

1. Pre-heat oven to 180 °C.   1. Heat on full power for 8-10 minutes.

2. Cook for 40 minutes.   2. Leave to stand for 2 minutes.

The graph shows the average energy efficiency of a typical
electric oven and a typical microwave oven.

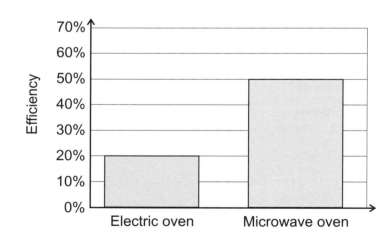

8 (b) (i) Describe the advantages of cooking the ready meal in a microwave oven instead of an electric oven.

....

....

....

....

....

....

....

[4]

8 (b) (ii) Microwave ovens use microwave radiation to cook food.
Give **two** other uses of microwave radiation.

1. ....

2. ....

[2]

8 (c) An electric oven is used for 1 hour a week. The power of the oven is 1.9 kW.

8 (c) (i) Calculate the amount of electrical energy, in joules, supplied to the oven each week.

....

....

....

Energy = ... J

[2]

8 (c) (ii) Electricity costs 10p per kWh.
Calculate how much it costs, in pence, to use the oven each week.

....

....

Cost = ... p

[1]

[Total 11 marks]

END OF QUESTIONS

Answers

Section One — Forces and Their Effects

Page 3: Speed and Velocity

1 a) A scalar quantity just has magnitude (size) *[1 mark]*.
   A vector quantity also has a direction *[1 mark]*.
   b) force *[1 mark]*
   c) 14 kg, 1 hour 9 minutes *[1 mark]*
2 a) has changed *[1 mark]*
   b) i) 8000 − 224 = **7776 m north** *[1 mark]*
      ii) average velocity = $s \div t$ = 7776 ÷ 1440 = **5.4 m/s north**
      [2 marks for correct answer, otherwise 1 mark for correct substitution into the formula.]

Page 4: Distance-Time Graphs

1 a) 160 s *[1 mark]*
   b) distance home = 480 m
   average velocity = $s \div t$ = 480 ÷ 160 = **3 m/s**
   [3 marks for correct answer, otherwise 1 mark for identifying the distance home and 1 mark for correct substitution into the formula.]
   c) The student did not run to training at a steady speed *[1 mark]* because the gradient for this part of the journey is not constant / because the line is not straight, and the gradient gives her speed *[1 mark]*.
   d) E.g.
   Distance from home in m

   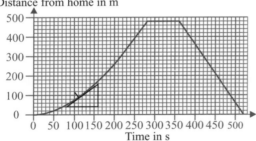

   velocity = $\dfrac{s}{t} = \dfrac{160 - 40}{160 - 80} = \dfrac{120}{80}$ = **1.5 m/s**

   [3 marks for an answer in the range 1.40 - 1.60 m/s, otherwise 1 mark for drawing a tangent at 110 s and 1 mark for attempting to divide a value of displacement by a value of time, both taken from the graph.]

Tangents can be a bit of a pain to draw, so make sure you draw them in pencil so you can rub them out if you need to. It can help if you draw a normal to the curve at the point you are interested in.

Pages 5-6: Acceleration and Velocity-Time Graphs

1 a) $a = \dfrac{v-u}{t} = \dfrac{20-0}{2.5}$ = **8 m/s²**
   [2 marks if answer correct, otherwise 1 mark for correct substitution into equation.]
   b) $a = \dfrac{v-u}{t} \Rightarrow v = (a \times t) + u = (8 \times 1.5) + 0$ = **12 m/s**
   [2 marks if answer correct, otherwise 1 mark for correct rearrangement of the equation and correct substitution of values. Full marks for a correct calculation using an incorrect answer from part 1 a).]
2 a) $a = \dfrac{v-u}{t} \Rightarrow u = v - (a \times t) = 5 - (0.5 \times 8)$ = **1 m/s**
   [2 marks if answer correct, otherwise 1 mark for correct rearrangement of the equation and substitution of values into the equation.]
   b) $a = \dfrac{v-u}{t} = \dfrac{8-5}{12}$ = **0.25 m/s²**
   [2 marks if answer correct, otherwise 1 mark for correct substitution of values into the equation.]

c) $a = \dfrac{v-u}{t} \Rightarrow t = \dfrac{v-u}{a} = \dfrac{0-7}{-3.5}$ = **2 s**
   [2 marks if answer correct, otherwise 1 mark for correct rearrangement of the equation and correctly substituting values into the equation.]
3 a) i) travelling at a steady velocity (20 m/s) *[1 mark]*
      ii) E.g. slowing down / (increasing) deceleration *[1 mark]*
   b) distance travelled = area under graph
      = (60 − 40) × (20 − 0)
      = **400 m**
      [3 marks for the correct answer, otherwise 1 mark for attempting to find the area under the graph between 40 and 60 seconds, 1 mark for correctly showing (60 − 40) × 20.]
   c) acceleration = gradient = $\dfrac{20-0}{40-0}$ = **0.5 m/s²**
      [3 marks for the correct answer, otherwise 1 mark for attempting to find the gradient, 1 mark for dividing a change in velocity over a change in time in the time range 0 − 40 s.]
   d) Velocity in m/s

   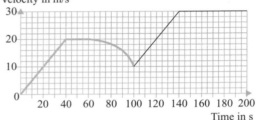

   [1 mark for a straight line with a positive gradient between 100 and 140 seconds, from 10 m/s to 30 m/s, 1 mark for a straight horizontal line between 140 and 200 seconds at 30 m/s.]

Page 7: Resultant Forces

1 a) The cyclist's weight is balanced by a (equal and opposite) reaction force from the saddle *[1 mark]*.
   b) A resultant force is a single force that has the same effect as all the individual forces acting on an object combined *[1 mark]*.
2 a) No — because the ball is not changing speed or direction / accelerating *[1 mark]*.
   b) Yes — because the ball is changing speed/accelerating *[1 mark]*.
   c) 5 N. The force exerted on the floor is equal to the force exerted by the ball on the floor, but in the opposite direction *[1 mark]*.

Page 8: Combining Forces

1 a) 6 000 000 − 1 500 000 = **4 500 000 N** *[1 mark]*
   b)

   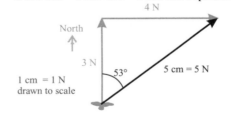

   Force = **5 N** *[1 mark]*
   Direction = **53°** (52-54° acceptable) *[1 mark]*

Finding the resultant force from a scale diagram is pretty straightforward. Just put the vectors end to end and join them up, then use a protractor to find the direction. And make sure you read the scale carefully.

2 a) Resultant force in the horizontal direction:
   17 + 3 − 10 − 10 = 0 N
   Resultant force in the vertical direction:
   10 − 2 = 8 N
   The resultant force on the kite is **8 N** *[1 mark]* **up** *[1 mark]*

b) $y - 4 = 0 \Rightarrow y = \textbf{4 N}$ *[1 mark]*

c) $20 - 5 - x = 0 \Rightarrow x = 20 - 5 = \textbf{15 N}$ *[1 mark]*

Pages 9-10: Forces and Acceleration

1 a) balanced *[1 mark]*

b) i) Zero / no resultant force *[1 mark]* because the van is travelling at a constant velocity / the van is not accelerating *[1 mark]*.

ii) $F = m \times a \Rightarrow a = \dfrac{F}{m} = \dfrac{200}{2500} = \textbf{0.08 m/s}^2$

[2 marks if answer correct, otherwise 1 mark for correct rearrangement of the equation and substitution of values into the equation.]

c) i) $F = m \times a = 10 \times 29 = \textbf{290 N}$

[2 marks if answer correct, otherwise 1 mark for correct substitution of values into the equation.]

ii) Force exerted on the van by the traffic cone = **290 N**

[1 mark if answer correct, or 1 mark for an answer that matches your answer to part ci).]

iii) Assuming all of this force causes the van to decelerate:

$F = m \times a \Rightarrow a = \dfrac{F}{m} = \dfrac{290}{2500} = \textbf{0.116 m/s}^2$

[2 marks if answer correct, otherwise 1 mark for correct rearrangement of the equation and correct substitution of values into the equation. Full marks for correct calculation using an incorrect answer from part c ii).]

2 a) The maximum force of the engine in each scooter ($= m \times a$)

$= 127.5 \times 2.4$

$= 306$ N

So, the mass of student B and her scooter is

$= \dfrac{F}{a} = \dfrac{306}{1.70} = \textbf{180 kg}$

[4 marks for the correct answer, otherwise 1 mark for correct substitution of mass and acceleration into the equation to find the driving force, 1 mark for correctly calculating F = 306 N and 1 mark for identifying that F ÷ 1.70 will give student B's mass.]

b) The scooter will accelerate more quickly / the acceleration will be higher when student C is riding it *[1 mark]*.

3 a) $F = m \times a = 75 \times 0.5 = \textbf{37.5 N}$

[2 marks for the correct answer, otherwise 1 mark for correct substitution into the equation.]

b) Driving force = resultant force – headwind

$= 37.5 - (-20) = \textbf{57.5 N}$

[2 marks for correct answer, otherwise 1 mark for identifying driving force as resultant force minus headwind.]

c) 730 N — because there will be an equal and opposite force from the road on the student (so there is no resultant force up or down) *[1 mark]*.

Page 11: Momentum and Collisions

1 a) $p = m \times v = 650 \times 15 = \textbf{9750 kg m/s}$

[2 marks if answer correct, otherwise 1 mark for correct substitution of values into the equation.]

b) momentum before = momentum after

$[m_1 \times v_1] + [m_2 \times v_2] = [(m_1 + m_2) \times v_{after}]$

$[9750] + [750 \times -10.2] = [(650 + 750) \times v_{after}]$

$9750 - 7650 = 1400 \times v_{after}$

$v_{after} = \dfrac{2100}{1400}$

$v_{after} = \textbf{1.5 m/s}$

[3 marks if answer correct, otherwise 1 mark for equating momentum before the collision and after, and 1 mark for correct rearrangement of the equation(s) and substitution of values into the equation. Allow full marks if an incorrect answer from part a) is used and the calculations are done correctly.]

2 Initial momentum of skater = $60 \times 5 = 300$ kg m/s

momentum of skater and bag = $(60 + \text{mass}_{bag}) \times 4.8$

momentum before = momentum after

$\Rightarrow 300 = (60 + \text{mass}_{bag}) \times 4.8$

$\Rightarrow \text{mass}_{bag} = \dfrac{300}{4.8} - 60 = \textbf{2.5 kg}$

[4 marks if answer correct, otherwise 1 mark for correct substitution of values into the equation for the momentum of the skater, 1 mark for correct substitution of values into the equation for the momentum of the skater with the bag, and 1 mark for equating the two and rearranging to find the mass of the bag.]

Page 12: Momentum and Safety

1 a) i) $p = m \times v = 1200 \times 30 = \textbf{36 000 kg m/s}$

[2 marks if answer correct, otherwise 1 mark for correct substitution of values into the equation.]

ii) $\text{force} = \dfrac{\text{change in momentum}}{\text{time taken}} = \dfrac{36\,000}{1.2} = \textbf{30 000 N}$

[2 marks if answer correct, otherwise 1 mark for correct substitution of values into the equation. Allow full marks if an incorrect answer from part a i) is used and the calculations are done correctly.]

b) i) The air bags increase the time taken by the driver to stop / change their velocity *[1 mark]*. This means the driver's momentum changes more slowly / the driver decelerates more slowly *[1 mark]*. This decreases the force acting on the driver *[1 mark]*.

ii) E.g. seat belts *[1 mark]*

c) $\text{force} = \dfrac{\text{change in momentum}}{\text{time taken}}$

Change in momentum = force × time taken

$= 4500 \times 5 = 22\,500$ kg m/s

Change in momentum = momentum before – momentum after $= m \times v_{before} - m \times v_{after}$

$m = \dfrac{\text{change in momentum}}{v_{before} - v_{after}} = \dfrac{22\,500}{27 - 12}$

$= \textbf{1500 kg}$

[3 marks for correct answer, otherwise 1 mark for correct calculation of change in momentum, and 1 mark for correct rearrangement of the equation and substitution into the equation to find mass.]

You could also have answered this question by calculating the car's deceleration and using F = ma. If you did this and got the correct answer, you still get full marks.

Pages 13-14: Frictional Force and Terminal Velocity

1 a)

[1 mark for an arrow pointing backwards.]

b) the same size *[1 mark]*

c) As the speed increases, the drag/resistive force increases *[1 mark]*.

2 a) i) This will ensure the weight and shape of the ball remain constant *[1 mark]*, so the only thing affecting the rate at which it falls is the parachute *[1 mark]*.

ii) E.g. dropping the ball from as large a height as possible *[1 mark]* to allow time to prepare for stopping the timer and to reduce uncertainty in the time reading *[1 mark]*. / Using parachutes with a much smaller mass than the steel ball *[1 mark]* so changing the parachute has little effect on the total mass *[1 mark]*. / Repeating each reading three (or more) times and calculating an average *[1 mark]* to reduce the chance of random errors or anomalous results affecting the outcome *[1 mark]*.

There are loads of possible answers you could give here — just make sure you answer the question properly and you explain your answer well.

b) When an object falls, resistive forces (e.g. air resistance) act on it in the opposite direction to its motion *[1 mark]*. These increase as the object accelerates *[1 mark]*. Eventually these resistive forces balance the downward force of an object's weight, so there is no longer a resultant force and the object falls at a steady velocity *[1 mark]*.

c) i)

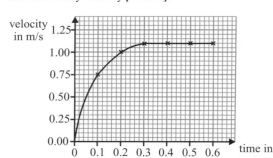

[3 marks available — 1 mark for 5 points plotted correctly or 2 marks for all points correctly plotted, and 1 mark for a suitable curved line of best fit]

ii) 1.1 m/s *[1 mark]*

A falling object has reached terminal velocity when it stops accelerating (i.e. its speed becomes constant). On a velocity time graph, it is the velocity corresponding to the flat part of the line of best fit.

d) Parachutes increase the drag /air resistance forces (at a given speed) *[1 mark]*. This means that the forces acting on the skydiver are balanced at a lower speed, so the terminal velocity is lower *[1 mark]*.

Page 15: Stopping Distances

1 a) i) The time it takes for a driver to respond to a hazard after noticing it *[1 mark]*.
   ii) E.g. tiredness *[1 mark]*, drug or alcohol intake *[1 mark]*
   b) it will decrease *[1 mark]*
   c) E.g. any three from: the speed of the car / the mass of the car / the condition of the road surface / the condition of the car's tyres / the weather conditions.
   [3 marks available — 1 mark for each correct answer.]
2 a) E.g. rain makes the road slippy and reduces friction between the tyres and the road *[1 mark]*.
   b) E.g. decrease their speed *[1 mark]*.

Page 16: Weight, Mass and Gravity

1 a) $W = m \times g \Rightarrow g = \dfrac{W}{m} = \dfrac{19.6}{2} = \textbf{9.8 N/kg}$

   [2 marks if answer correct, otherwise 1 mark for correct rearrangement of the equation and correct substitution of values into the equation.]
   b) 9.8 m/s² *[1 mark]*
   c) $W = m \times g = 2 \times 1.6 = \textbf{3.2 N}$
   [2 marks if answer correct otherwise 1 mark for correct substitution into the equation.]

Page 17: Work and Potential Energy

1 a) energy *[1 mark]*
   b) $W = F \times d = 50 \times 1600 = \textbf{80 000 J}$
   [2 marks for the correct answer, otherwise 1 mark for correctly substituting the values into the equation.]

2 a) i) $W = F \times d$ so $F = W \div d = 90\,000 \div 120 = \textbf{750 N}$
   [2 marks for the correct answer, otherwise 1 mark for correctly rearranging the equation and correctly substituting values into the equation.]
   ii) gravitational force / gravity *[1 mark]*
   b) The gravitational potential energy gained by the climber is equal to the work done = 90 kJ.

   $E_p = m \times g \times h$ so $m = \dfrac{E_p}{g \times h} = \dfrac{90\,000}{10 \times 120} = \textbf{75 kg}$

   [3 marks for correct answer, otherwise 1 mark for saying E_p = 90 kJ and 1 mark for correctly rearranging the equation and correctly substituting values into the equation.]

Pages 18-19: Kinetic Energy

1 a) movement. *[1 mark]*
   b) $E_k = \frac{1}{2} \times m \times v^2 = \frac{1}{2} \times 105 \times 3.0^2 = \textbf{472.5 J}$

   [2 marks for correct answer, otherwise 1 mark for correctly substituting values into the kinetic energy equation.]
   c) The friction between the brakes and the wheels *[1 mark]* transfers the kinetic energy of the cart into other forms (mostly heat) *[1 mark]*.

2 a) $E_k = \frac{1}{2} \times m \times v^2$ so $m = \dfrac{2E_k}{v^2} = \dfrac{2 \times 67\,500\,000}{15\,000^2} = \textbf{0.6 kg}$

   [2 marks for correct answer, otherwise 1 mark for correctly rearranging the equation and correctly substituting values into the equation.]
   b) E.g. Heat/light/sound *[1 mark]*. This energy transfer is caused by friction between the meteor and the Earth's atmosphere *[1 mark]*.

3 a) i) $E_k = \frac{1}{2} \times m \times v^2$ so, if the truck weighs twice as much as the car, it must have twice as much kinetic energy.
   $E_k = 2 \times 10\,000 = \textbf{20 000 J}$ *[1 mark]*.
   ii) $E_k = \frac{1}{2} \times m \times v^2$ so if the speed doubles, the kinetic energy will increase by a factor of $2^2 = 4$
   $E_k = 4 \times 10\,000 = \textbf{40 000 J}$ *[1 mark]*
   b) i) $E_k = \frac{1}{2} \times m \times v^2$ so

   $v = \sqrt{\dfrac{2E_k}{m}} = \sqrt{\dfrac{2 \times 432\,000}{1500}} = \textbf{24 m/s}$

   [2 marks for correct answer, otherwise 1 mark for correctly rearranging the equation and correctly substituting values into the equation.]
   ii) E_k = work done by brakes when stopping
   $E_k = F \times d$ so $d = E_k \div F = 432\,000 \div 7200 = \textbf{60 m}$
   [3 marks for correct answer, otherwise 1 mark for correctly equating kinetic energy of the car to the work done by the brakes and 1 mark for rearranging the equation $E_k = F \times d$ and correctly substituting values into the equation.]
   iii) The kinetic energy of the car increases with speed *[1 mark]*. The car has a maximum braking force, which means the car's minimum braking distance will also increase, so the driver must leave a larger distance between his car and the car in front to be able to stop safely *[1 mark]*.
   iv) increase *[1 mark]*

Pages 20-21: Forces and Elasticity

1 a) $e = 3$ cm $= 0.03$ m, $F = k \times e = 2 \times 0.03 = \textbf{0.06 N}$
   [2 marks for correct answer, otherwise 1 mark correctly substituting values into the equation.]
   b) 10 cm, the spring behaves elastically so it will return to its original length when the mass is removed from it.
   [1 mark for giving the correct length and reason.]

2 a) Her kinetic energy is transferred into elastic potential energy stored in the trampoline springs *[1 mark]*.

b) Force on each spring = $600 \div 30 = 20$ N
$F = k \times e$ so $k = F \div e = 20 \div 0.1 = \textbf{200 N/m}$
[3 marks for correct answer, otherwise 1 mark for calculating the force on each spring as 20 N, 1 mark for correctly rearranging the equation and correctly substituting values into the equation.]

3 a) E.g any two from: fixing the ruler with a clamp and marking the zero position of the end of the rope on the ruler / repeating each measurement at least three times and calculating averages / adding a separate marker to the show the top of the rope (to ensure it is at the same point each time). *[2 marks — one for each correct answer.]*

  b) i) limit of proportionality *[1 mark]*
   ii) Range = 0 N to 6.25 N *[1 mark]*

Page 22: Power

1 a) $P = W \div t = 7800 \div 60 = \textbf{130 W}$
[2 marks for correct answer, otherwise 1 mark for correctly substituting values into the equation.]

  b) i) $P = W \div t$ so $W = P \times t = 150 \times (10 \times 60) = \textbf{90 000 J}$
[2 marks for correct answer, otherwise 1 mark for correctly rearranging the equation and correctly substituting values into the equation.]
   ii) It will need recharging more often *[1 mark]*. The engine has a higher power and so will transfer more energy per second *[1 mark]*. It transfers chemical energy from the battery, and so you'd expect it to run the battery down/drain the battery more quickly *[1 mark]*.

2 Student A *[1 mark]*, because they transferred the most energy per second *[1 mark]*.
[1 mark for using the correct reasoning but giving an incorrect student due to a mistake in the calculations].

Pages 23-24: Turning Forces and Centre of Mass

1 a) i) The point on an object where all the mass of the object can be thought to be concentrated *[1 mark]*.
   ii)

[1 mark for a cross in the centre, by eye.]
If you're struggling with a question like this, have a go at drawing the shape's axes of symmetry. Handily the centre of mass of a symmetrical object always lies on its axes of symmetry. This means if an object has two or more axes of symmetry, it's easy to find the centre of mass — it must be where the lines cross.

  b) $M = F \times d = 20 \times 0.8 = \textbf{16 Nm}$
[2 marks for correct answer, otherwise 1 mark for correctly substituting values into the equation.]

2 a) B *[1 mark]* — the force is at the furthest distance from the pivot and is acting in a direction perpendicular to the handle *[1 mark]*.

  b) $M = F \times d$ so $d = M \div F = 0.675 \div 4.5 = \textbf{0.15 m}$
[2 marks for correct answer, otherwise 1 mark for correctly rearranging the equation and correctly substituting values into the equation.]

3 a) The centre of mass of the card, through which the object's weight can be thought to act, was directly below the point of suspension *[1 mark]*.

  b) i) Hang the plumb line from the same point as the piece of card and wait for it to come to rest *[1 mark]*. Draw a pencil line on the card along the plumb line *[1 mark]*. Hang the card from a different pivot and do the same thing again *[1 mark]*. Where the two lines cross is the centre of mass *[1 mark]*.
   ii) E.g. Make sure the card is not swaying when the lines are marked / is not moved by marking the lines / is not bent out of shape *[1 mark]*. / Make sure the line isn't too thick and is accurately placed *[1 mark]*.

Page 25: Balanced Moments and Levers

1 a) moment = force × perpendicular distance from the line of action to the pivot
$= 2 \times 0.2 = \textbf{0.4 Nm}$
[2 marks if answer correct, otherwise 1 mark for correct substitution of values into the equation.]

  b) clockwise moment about pivot = anticlockwise moments about pivot
$\text{force}_C \times \text{perpendicular distance}_C = 0.4 + 0.8$
$\text{perpendicular distance}_C = \dfrac{0.4 + 0.8}{8} = \textbf{0.15 m}$
[3 marks if answer correct, otherwise 1 mark for reference to balanced moments in each direction, and 1 mark for correct rearrangement of the equation and correct substitution of values into the equation. Allow full marks if an incorrect answer from part a) is used and the calculations are done correctly.]

2 E.g. Lifting a load by applying a force to the long handles means that there is a large distance between where the force is applied and the pivot *[1 mark]*. This means that a smaller force is needed to produce a particular moment in order to lift the load *[1 mark]*.

Page 26: Moments, Stability and Pendulums

1 a) The line of action of the weight lies inside the base of the cart *[1 mark]*.

  b) E.g. Having a wider base to lower the centre of mass *[1 mark]*. Having a heavier base to lower the centre of mass *[1 mark]*.

2 a) $T = 1 \div f = 1 \div 0.625 = \textbf{1.6 s}$
[2 marks for correct answer, otherwise 1 mark for correctly rearranging the equation and correctly substituting values into the equation.]

  b) $T = 1 \div f$ so $f = 1 \div T = 1 \div 1.25 = \textbf{0.8 Hz}$
[2 marks for correct answer, otherwise 1 mark for correctly rearranging the equation and correctly substituting values into the equation.]

  c) Change the length of the pendulum *[1 mark]*.

Page 27: Circular Motion

1 a) i) A centripetal force is the force that keeps an object moving in a circle *[1 mark]*.
   ii) tension (in the string) *[1 mark]*

  b) i) The ball is changing direction, so its velocity is changing. *[1 mark]*.
   ii) Towards the centre of the circle *[1 mark]*.

  c) The force needed would be greater *[1 mark]*.

  d) E.g. Increase the radius of the circle the ball travels in / increase the length of the string *[1 mark]*. / Decrease the speed of the ball *[1 mark]*.

Remember, the centripetal force of an object moving with circular motion increases as the speed or mass of the object increases, or as the radius of the circle the object moves in decreases.

Page 28: Hydraulics

1 a) incompressible *[1 mark]*

  b) E.g. hydraulic systems use two pistons — the load is placed on a piston with a large cross-sectional area and a force is applied on another piston with a smaller cross-sectional area to lift the load *[1 mark]*. When you apply a force to the small piston, it creates a pressure in the liquid between the two pistons ($P = F \div A$) *[1 mark]*. This pressure is transmitted equally through the liquid, so an equal pressure is created at the load piston *[1 mark]*. The load piston has a larger area — so the same pressure will create a much larger force on the load piston than the force applied to the smaller piston ($F = P \times A$) *[1 mark]*.

  c) i) $P = F \div A = 650 \div 0.0025 = \textbf{260 000 N/m}^2$ **(or Pa)**
[2 marks for the correct answer, otherwise 1 mark for correctly substituting the values into the equation.]

ii) The pressure created will be the same (260 000 N/m², or Pa) *[1 mark]* because pressure is transmitted equally through a liquid *[1 mark]*.

iii) $P = F \div A$ so $F = P \times A = 260\ 000 \times 0.09 =$ **23 400 N**
[2 marks for correct answer, otherwise 1 mark for correctly rearranging the equation and correctly substituting values into the equation.]

Section Two — Waves

Pages 29-30: Wave Basics

1 a) i) energy *[1 mark]*
   ii) matter *[1 mark]*
  b) i) R: rarefaction *[1 mark]*
       S: compression *[1 mark]*
   ii) A, because the oscillations are parallel to the direction of energy transfer / because it has rarefactions and compressions *[1 mark]*.
   iii) The direction of oscillation is perpendicular to the direction of energy transfer *[1 mark]*.
  c) i) E.g. sound waves / shock waves / ultrasound *[1 mark]*
   ii) E.g. light (or any type of electromagnetic wave) / waves on string / ripples on water *[1 mark]*.

2 a) 2 Hz *[1 mark]*
  b) $v = f \times \lambda$ so $\lambda = \dfrac{v}{f} = \dfrac{0.5}{2} =$ **0.25 m**
     [2 marks for the correct answer — otherwise 1 mark for correctly rearranging the equation and substituting the correct values into the equation.]
     Remember, the wavelength of a wave is the distance from crest to crest.
  c) In one second, there are two complete waves, so it takes $1 \div 2 = 0.5$ s for one complete wave, so the time period is **0.5 s** *[1 mark]*.

3 a) i) The amplitude is the displacement of the wave from its rest position to a crest *[1 mark]*.
   ii) A and C *[1 mark]*
  b) E.g.

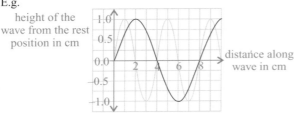

height of the wave from the rest position in cm / distance along wave in cm

[1 mark for a wave with an amplitude of 1 cm and a wavelength of 8 cm.]

Page 31: Reflection of Waves

1 a) i) 20° *[1 mark]*
   ii) An imaginary line that is at right angles to the surface at the point where the ray of light hits the surface *[1 mark]*.
  b) i)

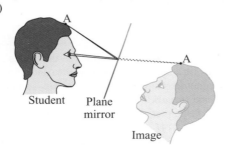

Student   Plane mirror

Image

[2 marks available — 1 mark for reflected rays drawn from point A on the student to the student's eyes, and 1 mark for virtual rays drawn from the point of reflection on the mirror to point A in the image. OR 1 mark for each correctly drawn ray from point A that is reflected in the mirror and ends at the eye.]
   ii) virtual *[1 mark]*

Page 32: Diffraction and Interference

1 a) Waves bending and spreading out as they travel past edges or through gaps *[1 mark]*.
  b) i) Visible light has a very short wavelength compared to a doorway *[1 mark]* so its diffraction is too small to notice *[1 mark]*.
     Visible light does diffract a very small amount, but it's too small for us to notice.
   ii) Their wavelength is roughly the same size as the width of a doorway, so they will diffract noticeably *[1 mark]*.

2 a) Interference / they will interfere with each other *[1 mark]*.
  b)

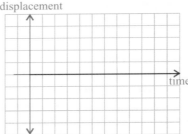

displacement / time

[1 mark for a straight line along the x-axis.]
When the waves meet, they cancel each other out, so there is no resulting sound wave and the air doesn't move.

Page 33: Refraction of Waves

1 a) The two materials have different densities and so the ray travels at different speeds in them *[1 mark]*. The ray is incident at the boundary between the two materials at an angle to the normal, which means it will change direction as it changes speed *[1 mark]*.
  b) B. The light refracts (bends) towards the normal as it crosses the boundary *[1 mark]*.
  c) The light would not refract (bend) *[1 mark]*.

2 a) E.g.

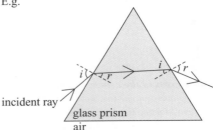

incident ray / glass prism / air

[3 marks available — 1 mark for refracting the ray towards the normal upon entering the prism, 1 mark for refracting the ray away from the normal as it leaves the prism and 1 mark for correctly labelling all the angles of incidence and refraction.]
  b) E.g. Place the prism on a piece of paper and shine a ray of light at the prism. Trace the incident and emergent rays and the boundaries of the prism on the piece of paper *[1 mark]*. Remove the prism and draw in the refracted ray through the prism by joining the ends of the other two rays with a straight line *[1 mark]*. Draw in the normals using a protractor and a ruler *[1 mark]* and use the protractor to measure i and r *[1 mark]*.
  c) dispersion *[1 mark]*

Page 34: Refractive Index

1 a) Refractive index $= \dfrac{\text{speed of light in a vacuum or air}}{\text{speed of light in the medium}}$
       $= \dfrac{300\ 000\ 000}{200\ 000\ 000} =$ **1.5**
     [2 marks for correct answer, otherwise 1 mark for correct substitution of values into the formula.]

b) Angle of incidence for violet light = $i = 45$ °

$$\sin r = \frac{\sin i}{n} = \frac{\sin 45°}{1.528} = 0.4627...$$

$$\Rightarrow r = \sin^{-1}(0.4627...) = 27.565....°$$

Then subtract this angle from the angle of refraction of red light to get:

$\theta = 28.13 - 27.5765... = 0.5597...° = $ **0.56 ° (to 2 s.f.)**

[4 marks for correct answer, otherwise 1 mark for correct angle of refraction of violet light, 1 mark for correct substitution and rearrangement of the equation to find r and 1 mark for correctly subtracting one angle from the other.]

2 $n = \frac{\sin i}{\sin r}$ so $\sin i = (n \times \sin r) = 0.444626...$

$i = \sin^{-1}(0.444626...) = 26.3994... = $ **26.40 ° (to 4 s.f.)**

[2 marks for a correct answer, otherwise 1 mark for correctly rearranging the equation and correct substitution into the equation.]

Page 35: Total Internal Reflection

1 a) Total internal reflection is when light is completely reflected at the boundary between media *[1 mark]*. It occurs when light hits a boundary with a less dense medium at an angle greater than the critical angle *[1 mark]*.

b) E.g. endoscopes *[1 mark]*

2 a) The angle of incidence at a boundary at which the angle of refraction is 90°. / The angle of incidence at a boundary at which all of the light is refracted along the boundary between the media *[1 mark]*.

b) less than *[1 mark]*.

c) It will be reflected back into the acrylic (total internal reflection) *[1 mark]*.

d) $n = \frac{1}{\sin c} = \frac{1}{\sin 41.8°} = 1.50030.... = $ **1.50 (to 3 s.f.)**

[2 marks for correct answer, otherwise 1 mark for substituting correct values into the equation.]

Pages 36-38: Electromagnetic Waves and Their Uses

1 a) ultraviolet *[1 mark]*

b) radio waves *[1 mark]*

c) gamma rays *[1 mark]*

d) at the same speed as *[1 mark]*

e) 10^{-15} m $- 10^4$ m *[1 mark]*

2 a) i) Long-wave radio signals can diffract round the mountain and reach the house *[1 mark]* because they have a long wavelength *[1 mark]*.

ii) E.g. television signals / Bluetooth® *[1 mark]*

b) microwave (radiation) *[1 mark]*

c) infrared *[1 mark]*

3 Valuables are marked with a special/fluorescent ink *[1 mark]*. This ink is invisible under normal light, but glows/fluoresces/becomes visible under UV light *[1 mark]*.

4 a) microwaves *[1 mark]*

b) radio waves *[1 mark]*

c) i) Any two of: e.g.
Always use the same picture *[1 mark]* because different pictures may be easier/harder to transmit *[1 mark]*. / Always use the same phone/laptop *[1 mark]* because different phones/laptops may have different ranges *[1 mark]*. / Always do the test in the same place *[1 mark]* in case other devices produce signals which interfere with the Bluetooth® signals / in case materials/obstacles/walls in the area interfere with the Bluetooth® signals *[1 mark]*. / Always make sure the phone and the Bluetooth® receiver on the laptop are pointing towards each other *[1 mark]* because changing the direction they face in might change how far the signal has to travel / could cause signal loss *[1 mark]*.

There are lots of possible answers here, just make sure you have two sensible suggestions, and a reason for each.

ii) E.g. the signal is being absorbed/reflected/blocked by the door/wall *[1 mark]*.

d) E.g. optical fibres *[1 mark]*

5 a) gamma rays *[1 mark]*

b) Treating the fruit with radiation kills the microbes/bacteria in it *[1 mark]*, which means that it will stay fresh for longer *[1 mark]*.

Page 39: The Dangers of Electromagnetic Waves

1 a) E.g. (skin) cancer *[1 mark]*, blindness *[1 mark]*

b) E.g. genetic mutations *[1 mark]*

2 X-rays are ionising, which means they can knock electrons from atoms in cells *[1 mark]*. This can cause cell damage/cell death/mutation/cancer *[1 mark]*.

3 a) Microwaves can cause internal heating of human body tissue *[1 mark]*.

b) Infrared has a higher frequency/carries more energy than microwaves so it's more likely to cause burns *[1 mark]*.

Page 40: X-Rays

1 X-rays are ionising *[1 mark]*. These means they can kill the cancer cells *[1 mark]*.

2 a) high *[1 mark]*

b) X-rays are absorbed by hard tissue (like bone) *[1 mark]*, but transmitted by soft tissue *[1 mark]*. The film turns dark in areas where X-rays can get through, but stays light in areas where they can't *[1 mark]*.

The X-rays interact with the photographic film in the same way as light.

c) CT (computed tomography) scans *[1 mark]*

d) E.g. fractured bones / dental problems *[1 mark]*

e) i) E.g. wear a lead apron *[1 mark]* / stand behind a lead screen *[1 mark]* / leave the room *[1 mark]*.

ii) E.g. minimise exposure time *[1 mark]* / avoid repeatedly taking X-rays of the same part of the patient *[1 mark]* / use lead to shield other areas of the body not being imaged *[1 mark]*.

Page 41: Sound and Ultrasound

1 a) A *[1 mark]*

b) C *[1 mark]*

The higher the frequency, the higher the pitch of the sound.

c) i) 20 Hz to 20 000 Hz *[1 mark]*

ii) By a computer / electronically *[1 mark]*.

d) Because sound waves travel as vibrations in a medium and in a vacuum there is no medium to vibrate *[1 mark]*.

2 a) Echoes are caused by the reflection of sound waves from the surfaces of the drama hall, but on the field there are no walls for the waves to reflect off *[1 mark]*.

b) The sound waves diffract (spread out) as they pass through the doorway of the hall *[1 mark]*.

Page 42: More on Ultrasound

1 a) An ultrasound pulse is emitted by a computer into the area of the body being scanned. At the boundary between two different media the wave is partially reflected *[1 mark]*. These reflected waves are detected and used to generate an image *[1 mark]*.

b) E.g. treating kidney stones *[1 mark]*. High-energy waves are targeted at the stones, breaking them up into small particles *[1 mark]*.

2 $t = 5 \times 0.000005 = 0.000025$ s
$s = v \times t = 1450 \times 0.000025 = 0.03625$ m
The thickness is half this distance, so
$0.03625 \div 2 = 0.018125$ m = **18.125 mm**
[2 marks if correct, otherwise one mark for correctly reading the time from the diagram and correctly substituting this into the equation.]

Don't get caught out on ultrasound questions — typically the time you are given will be the time for the sound to travel from the receiver to the boundary and back again.

Page 43: Lenses and Magnification

1 a) converging *[1 mark]*

b) Light is refracted towards the axis of the lens as it enters the lens and again as it exits *[1 mark]*. / Each point along the lens acts like a small prism, bending light towards the axis *[1 mark]*.

c) The point on the lens axis where rays of light entering the lens parallel to the axis are brought to a focus *[1 mark]*.

d) The distance between the centre of the lens and the principal focus *[1 mark]*.

e) In a real image, light converges/comes together to form an image at the point where it is seen *[1 mark]*. In a virtual image, light appears to come from a point it has not actually been through *[1 mark]*.

f) size *[1 mark]*, orientation / which way up it is *[1 mark]*

g) magnification = $\dfrac{\text{image height}}{\text{object height}}$

object height = $\dfrac{\text{image height}}{\text{magnification}} = \dfrac{10}{2.5} = \textbf{4 mm}$

[2 marks for the correct answer, otherwise 1 mark for correctly rearranging the formula and substituting the correct values.]

Pages 44-45: Converging Lenses

1 a) Any ray passing through the centre of the lens *[1 mark]*.

b)

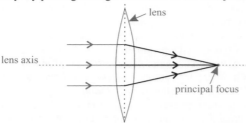

[2 marks for a correct answer — 1 mark for top and bottom rays bending towards the axis at the centre of the lens, or once on entering the lens and again on leaving the lens. 1 mark for the three rays meeting at the principal focus.]

c) E.g.

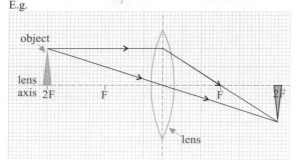

[4 marks for a correct answer — 1 mark for an inverted image of the triangle below the axis at 2F, 1 mark for drawing the image the same size as the object. 1 mark for each of two correctly drawn rays from the object passing through the lens to the image.]

2 a) E.g. Place the lens in front of a screen, and angle both towards a distant object *[1 mark]*. Move the lens away from the screen until a focussed image of the object appears on the screen *[1 mark]*. Measure the distance between the lens and the screen. This is the focal length *[1 mark]*.

There is more than one way to measure the focal length of a lens. If you have been taught a different method in school and can describe it fully you will still get full marks.

b) To make her answer more reliable / to identify any anomalous results *[1 mark]*.

3 a) Convex *[1 mark]*

b) The image is the right way up *[1 mark]*, virtual *[1 mark]*, and larger than the ladybird *[1 mark]*.

c) The image becomes a real image *[1 mark]*, and flips upside down *[1 mark]*.

Page 46: Diverging Lenses

1 a) a concave lens *[1 mark]*

b) The image is always a virtual image *[1 mark]*. / The image is always smaller than the object *[1 mark]*. / The image is always the right way up *[1 mark]*.

c)

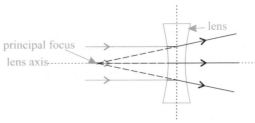

[1 mark for top and bottom rays bending away from the axis at the centre of the lens. 1 mark for projecting the refracted rays back to the principal focus.]

d)

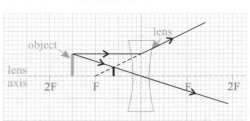

[3 marks for a correct answer — 1 mark for drawing a ray from the top of the object passing through the centre of the lens without refracting. 1 mark for drawing a ray travelling from the top of the object parallel to the lens axis and refracting away from the axis at the centre of the lens so that it appears to have come from the principle focus. 1 mark for placing the top of the image at the point where the ray travelling through the centre of the lens meets the path the refracted ray appears to have taken from the principle focus.]

Page 47: Power and the Lens Equation

1 a) 40 cm = 0.4 m

power = $\dfrac{1}{f} = \dfrac{1}{0.4} = \textbf{2.5 D}$ *[1 mark]*

Whenever you're substituting values into an equation make sure your units are correct.

b) power = $\dfrac{1}{f} = 20$

The lens equation is: $\dfrac{1}{f} = \dfrac{1}{u} + \dfrac{1}{v}$

So $20 = \dfrac{1}{u} + \dfrac{1}{0.3}$

$\dfrac{1}{u} = 20 - \dfrac{1}{0.3} = 16.666666666...$

$u = 1 \div 16.6666666... = 0.06 \text{ m} = \textbf{6 cm}$

[3 marks for a correct answer, otherwise 1 mark for correctly calculating the focal length or showing $1 \div f = 20$, and 1 mark for correctly substituting in the values of f or $1 \div f$ and v and correctly rearranging the lens equation to find u.]

c) Virtual, because the object is less than one focal length away from the lens *[1 mark]*.

d) E.g. lens A could be made from a material with a higher refractive index than lens B *[1 mark]*. Lens A could be more curved than lens B *[1 mark]*.

Pages 48-49: The Eye

1 a) D: suspensory ligaments *[1 mark]*
E: retina *[1 mark]*
F: lens *[1 mark]*

b) i) the iris *[1 mark]*

The iris controls the size of the pupil, the hole through which light enters the eye.

ii) the retina *[1 mark]*

iii) The cornea is the transparent convex layer of the eye *[1 mark]* that has a high refractive index *[1 mark]*. The cornea refracts light as it passes into the eye (and does most of the eye's focusing) *[1 mark]*.

2 a) The near point is the closest distance at which an eye can focus *[1 mark]*.

b) infinity *[1 mark]*

c) range *[1 mark]*

d) The ciliary muscles control the thickness of the eye's lens *[1 mark]*. When the ciliary muscles relax, the lens is stretched *[1 mark]*. When the ciliary muscles contract, the lens becomes thicker *[1 mark]*. Changing the shape of the lens changes its focal length, meaning the eye is able to focus on objects at different distances *[1 mark]*.

3 E.g. Both the eye and a camera make a real, inverted image by focusing light through a lens *[1 mark]*. The retina of the eye records the light that falls on it as the film or CCD does in a camera *[1 mark]*. In the eye, light from different distances is focused by the lens changing shape, and in a camera light is focused by the lens moving relative to the film *[1 mark]*. The pupil of the eye can get bigger or smaller to control how much light comes in. A camera controls light levels by changing the size of the aperture *[1 mark]*.

Page 50: Correcting Vision

1 a) short sight *[1 mark]*

b) The eyeball may be too long *[1 mark]*. The cornea/lens may be too powerful/refracting light too much/the eye lens is unable to focus *[1 mark]*.

c) i) A laser is a concentrated source/beam of light *[1 mark]*.

ii) Laser eye surgery can change the shape of the cornea to help the eye focus light *[1 mark]*.

2 a) convex/converging *[1 mark]*

b) In long sight, the eye is unable to focus on near objects as it cannot refract light strongly enough *[1 mark]*. Wearing glasses with convex lenses will mean light starts to converge before it enters the eye *[1 mark]*. This will mean that the eye will now be able to refract light enough to be able to focus it *[1 mark]* and form an image on the retina *[1 mark]*.

Page 51: The Doppler Effect and Red-Shift

1 a) i) it decreases *[1 mark]*

ii) the Doppler effect *[1 mark]*

b) As the ambulance approaches, the wavelength of the sound of the siren appears shorter *[1 mark]* so the (frequency and therefore the) pitch is higher *[1 mark]*. When the ambulance starts moving away, the wavelengths of the sound appear to become longer *[1 mark]*, so the pitch is lower *[1 mark]*.

c) E.g visible light / microwaves *[1 mark]*

2 a) The increase observed in the wavelength of light from distant galaxies *[1 mark]*.

b) Red-shift observations show that distant galaxies are moving away from us *[1 mark]*. The more distant the galaxy, the faster it appears to be moving away from us *[1 mark]*.

Page 52: The Big Bang

1 a) i) The Big Bang theory states that the universe began from a very small point, which exploded and started to expand *[1 mark]*.

ii) It suggests the universe is expanding, which supports the idea that in the past it was smaller and could have started from a small point *[1 mark]*.

b) i) Cosmic microwave background radiation is low frequency electromagnetic radiation *[1 mark]* that we can detect coming from every direction in space *[1 mark]*.

ii) Shortly after the Big Bang, the whole universe emitted radiation that we now observe as cosmic microwave background radiation *[1 mark]*.

c) E.g. It is the currently accepted theory for how the universe began and has a lot of evidence supporting it, but no theory is ever definitely true *[1 mark]*.

Section Three — Heating Processes

Page 53: Kinetic Theory and Changes of State

1 a) i) X – gas, Y – liquid
[2 marks available — 1 mark for each correct answer.]

ii) gas(es) *[1 mark]*

b) i) When a solid is heated the particles inside it gain energy and vibrate faster *[1 mark]*. When the temperature gets high enough, they have enough energy to overcome the forces of attraction between them and start moving around (i.e. the substance becomes a liquid) *[1 mark]*.

ii) The graph is flat showing the temperature isn't changing *[1 mark]*. This is because the wax is changing from a solid to a liquid *[1 mark]*, so the energy supplied is being used to break intermolecular bonds rather than raising the temperature of the wax *[1 mark]*.

iii) 15 minutes *[1 mark]*

After 15 minutes, the wax starts getting hotter again, which means the energy is now being used to increase its temperature rather than breaking intermolecular bonds.

iv) E.g. it has a different amount of impurities *[1 mark]*.

Page 54: Specific Heat Capacity

1 a) i) The amount of energy needed to raise the temperature of 1 kg of the substance by 1°C *[1 mark]*.

ii) Mercury *[1 mark]* as it has the lowest specific heat capacity *[1 mark]*.

b) Rearrange the equation for specific heat capacity:
$m = E \div (c \times \theta)$.
$m = 3040 \div (380 \times 40) = 3040 \div 15\,200$
$= \mathbf{0.2\ kg}$
[2 marks for correct answer, otherwise 1 mark for substituting into the correctly rearranged formula.]

c) $E = m \times c \times \theta$.
The temperature change for both is 50 °C.
Energy from mercury = $27.2 \times 139 \times 50 = 189\,040$ J.
Energy from water = $2.0 \times 4200 \times 50 = 420\,000$ J.
Difference = $420\,000 - 189\,040 = \mathbf{230\,960\ J}$
[3 marks for correct answer, otherwise 1 mark substituting into the correct formula for mercury and for water and 1 mark for attempting to calculate the difference by subtracting the energy released by the mercury from the energy released by the water.]

Page 55: Specific Latent Heat

1 a) i) The amount of energy needed to change 1 kg of the substance from a solid to a liquid without changing its temperature *[1 mark]*.

ii) Copper *[1 mark]*, because it has the highest specific latent heat of vaporisation *[1 mark]*.

b) $E = m \times L_f$, rearranged gives:
$m = E \div L_f$
$= 184 \div 23 = \mathbf{8\ kg}$
[2 marks for correct answer, otherwise 1 mark for correctly substituting into the correctly rearranged equation.]

c) i) 30 g of ice completely melts.
30 g = 0.03 kg
L_f of water = 334 kJ/kg = 334 000 J/kg
$E = m \times L_f = 0.03 \times 334\,000 = \mathbf{10\,020\ J}$
[2 marks for correct answer, otherwise 1 mark for correctly substituting into the correct equation.]

ii) $E = m \times L_v = 0.15 \times 2\,260\,000 = 339\,000$ J
The heat source supplies 250 J every second, so would need to be on for $339\,000 \div 250 = \mathbf{1356\ s}$.
[3 marks for correct answer, otherwise 1 mark for correctly substituting into the correct equation for energy and 1 mark for dividing the energy by the energy transferred by the heat source per second.]

Page 56: Heat Radiation

1 a) Flask C is made from a dark matt material, which is a good emitter of infrared radiation *[1 mark]* whereas flask B is made from a light, shiny material which is a poor emitter of infrared radiation *[1 mark]*.

b) i) They will all absorb radiation from the surroundings *[1 mark]*, because all objects absorb (and emit) infrared radiation *[1 mark]*.
The flasks cool down because they emit more radiation than they absorb, not because they don't absorb any radiation at all.

ii) B, because it is hotter so will radiate the most radiation *[1 mark]*.

c) There is a larger temperature difference between flask B and the ice bath *[1 mark]*.

d) The shiny, white surface of flask A reflects infrared radiation *[1 mark]* better than the matt black surface of flask B *[1 mark]*. / The matt black surface of flask C is a better absorber *[1 mark]* of infrared radiation than the shiny white surface of flask A *[1 mark]*.

2 A dark, matt surface *[1 mark]* because this will make the panel a good absorber of infrared radiation so it will absorb a lot of the Sun's energy *[1 mark]*.

Page 57: Conduction and Convection

1 a) solid *[1 mark]*

b) The particles in a solid are closer together than in a liquid *[1 mark]*, so the kinetic energy can be passed between particles more quickly and easily *[1 mark]*.

c) i) E.g.

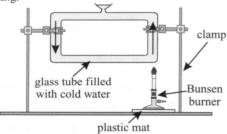

clamp

glass tube filled with cold water

Bunsen burner

plastic mat

[1 mark for two arrows drawn anywhere inside the glass tube showing the correct flow of water.]

ii) The water particles near the heat source gain energy *[1 mark]*. This causes the water near the heat source to expand and become less dense and rise up the pipe *[1 mark]*. Colder, denser water elsewhere in the pipe is displaced and moves to replace this heated water *[1 mark]*.

d) i) The particles nearest the Bunsen burner will gain kinetic energy and vibrate faster *[1 mark]*. They then collide with their neighbouring particles, transferring some of their extra kinetic energy to them *[1 mark]*.

ii) Metals are good conductors because they contain free electrons that transfer energy more quickly throughout the metal *[1 mark]*.

Page 58: Condensation and Evaporation

1 When water vapour comes into contact with a cold car window, it transfers energy to the window *[1 mark]*. As this happens, the particles in the water vapour lose kinetic energy and are pulled closer together by forces of attraction *[1 mark]*, turning it from a gas into a liquid *[1 mark]*

2 a) How to grade your answer:
0 marks: There is no relevant information.
1-2 marks: There is a brief mention of how the evaporation of sweat cools you down, with no real explanation.
3-4 marks: There is an explanation of how the evaporation of sweat can cool you down, with reference to particles.
5-6 marks: There is a clear and detailed explanation of how the evaporation of sweat cools you down, in terms of the movement of particles.
Here are some points your answer may include:
Particles in the liquid sweat evaporate.
Only the fastest particles with the most kinetic energy evaporate from the liquid.
The slower particles are left in the liquid.
The average kinetic energy of the particles in the liquid sweat decreases.
This means the average temperature of the sweat decreases, cooling the skin.

b) less humid *[1 mark]*

c) Greater airflow over a liquid increases its rate of evaporation *[1 mark]*. This is because the air above the liquid is replaced quicker, so the concentration of sweat in the air just above the sweat layer will be lower than if there was less wind *[1 mark]*.

Page 59: Rate of Heat Transfer and Expansion

1 a) Fox B. Fox B has large ears which means they have a large surface area *[1 mark]*. This means they lose heat more easily (by radiation) *[1 mark]*.

b) The longer hairs trap a thicker layer of insulating air around the body *[1 mark]*, limiting the amount of heat lost to its surroundings by convection *[1 mark]*.

2 a) E.g. the cooling fin may have a large surface area *[1 mark]*, so it can lose heat to its surroundings by radiation quickly *[1 mark]*. The cooling fin may be made of metal *[1 mark]* so heat will be conducted across the fin quickly *[1 mark]*.

b) The gel is a better conductor than air, so heat will be transferred to the fin more quickly *[1 mark]*.

c) i) When the temperature rises, the metals expand *[1 mark]*. They are joined together and expand by different amounts, causing the strip to bend *[1 mark]*.

ii) E.g. roofs/bridges may expand and collapse/become dangerous *[1 mark]*.

Page 60: Energy Transfer and Efficiency

1 a) i) Energy can be created. *[1 mark]* Energy cannot be transferred. *[1 mark]*

ii) 20 J of energy *[1 mark]* — energy is conserved, so the total output energy must equal the total input energy *[1 mark]*.
The total output energy is 8 J + 11.5 J + 0.5 J = 20 J.

b) i) 8 J *[1 mark]*

ii) efficiency $= \dfrac{\text{useful energy out}}{\text{total energy in}} = \dfrac{8}{20} = \mathbf{0.4\ (or\ 40\%)}$
[2 marks if answer correct, otherwise 1 mark for correct substitution of values into the equation. Receive 2 marks for an incorrect answer calculated correctly using an incorrect answer (or answers) from part aii) or bi).]

c) i) efficiency $= \dfrac{\text{useful power out}}{\text{total power in}}$
so useful power out = efficiency × total power in
$= 0.75 \times 20 = \mathbf{15\ W}$
[2 marks if answer correct, otherwise 1 mark for correct substitution of values into the correctly rearranged equation.]

ii) It will be transferred/dissipated to the surroundings as heat *[1 mark]*.

Pages 61-62: Sankey Diagrams

1 a) i) 10 J *[1 mark]*
You know the total input energy is 200 J. The input energy arrow is 20 squares wide, so the value of each square must be 200 J ÷ 20 = 10 J.

  ii) 50 J *[1 mark]*
The useful energy arrow is 5 squares wide, and each square represents 10 J. So the amount of energy that's usefully transferred = 5 × 10 J = 50 J.

  iii) E.g.

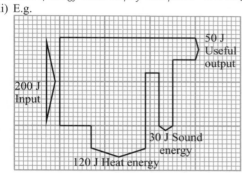

  [3 marks available — 1 mark for three arrow widths drawn correctly, 1 mark for all four arrow widths drawn correctly, 1 mark for all arrows being correctly labelled.]

 b) i) Gravitational potential energy of lifted weight
    = 100 − 50 − 20 = 30 kJ *[1 mark]*

  ii) E.g.

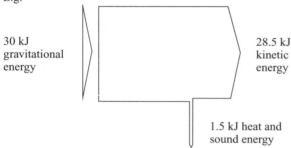

  [3 marks available — 1 mark for drawing a recognisable Sankey diagram, 1 mark for all of the arrows being drawn in roughly the correct proportions, 1 mark for all of the arrows being correctly labelled. Receive a mark if the arrows are labelled incorrectly due to an incorrect answer for part bi).]

Page 63: Energy Efficiency in the Home

1 a) i) conduction *[1 mark]*

  ii) E.g. it will help stop a convection current being set up in the air gap (and so reduce heat loss by convection). / It will reduce radiation across the gap (as the foam will reflect/absorb radiation).
    [1 mark for naming a type of heat transfer and correctly explaining how energy transfer is reduced].

 b) i) 60 years *[1 mark]*
This is the initial cost divided by the annual saving — 1200 ÷ 20 = 60.

  ii) The window shutters have a longer payback time, so they're less cost-effective in the long run *[1 mark]*.

  iii) E.g. Fitting draught-proofing strips around the windows *[1 mark]* will help to reduce the amount of hot air escaping from the house and cold air coming in the house through any gaps, reducing heat loss by convection *[1 mark]*. / Putting up curtains/buying thicker curtains *[1 mark]* — curtains will help to reduce the amount of warm air in a room reaching the window, and so reduce heat loss by conduction/convection *[1 mark]*. / Putting up curtains/buying thicker curtains *[1 mark]* — curtains will absorb/reflect heat radiation that would otherwise escape through the window *[1 mark]*.

 c) She should choose brand A because it has a lower U-value / is a better insulator *[1 mark]*.

Section Four — Electricity

Page 64: Current and Potential Difference

1 a) i) $I = \frac{Q}{t}$, so $Q = I \times t = 5 \times (20 \times 60) = $ **6000 C**
    [2 marks for correct answer, otherwise 1 mark for correctly substituting into the rearranged equation.]

  ii) $V = \frac{E}{Q}$ so $E = V \times Q = 3 \times 6000 = $ **18 000 J**
    [2 marks for correct answer, otherwise 1 mark for correctly substituting into the rearranged equation. Allow marks if an incorrect value from a) i) is used correctly.]

 b) i) Electric current *[1 mark]*.

  ii) E.g. copper *[1 mark — accept metal]*, because electrical charges can flow through it easily *[1 mark]*.

 c) The reading will be much lower / will be decreased *[1 mark]*.

Page 65: Circuits — The Basics

1 a) ⊣▭⊢ *[1 mark]*

 b) i) B — cell *[1 mark]*
    C — variable resistor *[1 mark]*

  ii)

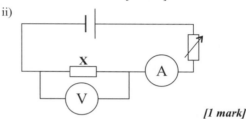

    [1 mark]

 c) How to grade your answer:
    0 marks: There is no relevant information.
    1-2 marks: There is a brief description of an experimental method.
    3-4 marks: There is a clear description of an experimental method.
    5-6 marks: There is a clear and detailed description of an experimental method.
    Here are some points your answer may include:
    Fix the resistance of the variable resistor so that the potential difference is fixed at a certain value.
    Measure the current using the ammeter.
    Measure the potential difference using the voltmeter.
    Change the resistance of the variable resistor until you get a certain value of potential difference and take another reading of current.
    Take several pairs of readings.
    Take repeat readings of I for each value of V and average.
    Turn off the circuit between each reading to avoid heating, which will cause the current to decrease.

 d) Resistance *[1 mark]*

Pages 66-67: Resistance and V = I × R

1 a) resistor *[1 mark]*

 b) D *[1 mark]*
The graph with the shallowest gradient corresponds to the component with the highest resistance.

 c) i) $V = I \times R$ so $R = V \div I$
    The resistance is 1 ÷ gradient and the gradient is constant (it's a straight-line graph), so the resistance is constant.
    So take the point (2, 4).
    $R = \frac{V}{I} = \frac{2}{4} = $ **0.5 Ω**
    [3 marks for correct answer, otherwise 1 mark for saying a constant gradient means constant resistance, or for calculating the gradient, and 1 mark for correctly substituting values into the resistance equation, or for calculating 1/gradient.]

  ii) $V = I \times R$ so $I = \frac{V}{R} = \frac{15}{0.75} = $ **20 A**
    [2 marks for correct answer, otherwise 1 mark for correctly substituting into the correct rearranged equation.]

2 a) A diode *[1 mark]* because current only flows in one direction *[1 mark]*.

b)

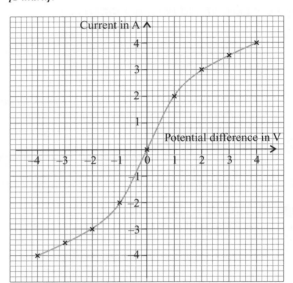

[3 marks available — 1 mark for 8 points plotted correctly or 2 marks for all points correctly plotted, and 1 mark for a correct curved line of best fit.]

c) i) (filament) lamp / bulb *[1 mark]*

ii) As the current through the component/filament increases, the temperature increases *[1 mark]*. The ions in the component/filament gain heat energy and start to vibrate more *[1 mark]*. It is harder for charge-carrying electrons to pass through the component/filament, so its resistance increases *[1 mark]*. So the value of V ÷ I increases, causing the graph to curve *[1 mark]*.

Page 68: Circuit Devices

1 a i) The current increases *[1 mark]*. As the temperature in the room increases, the resistance of the thermistor (and so the circuit) will decrease *[1 mark]*.

ii) The LED would light up *[1 mark]*.

iii) E.g. thermostat *[1 mark]*

iv) When an a.c. electricity supply is only able to flow in one direction through the circuit *[1 mark]*, so only half the cycle can flow *[1 mark]*.

b) i) E.g. the resistance of a thermistor is affected by temperature whereas the resistance of an LDR is affected by light intensity *[1 mark]*.

ii) E.g. to control automatic night lights *[1 mark]*.

iii) E.g. they use a much higher current than LEDs *[1 mark]*.

Pages 69-70: Series Circuits

1 a) i) E.g. if one fairy light breaks, the rest won't light up *[1 mark]*.

ii) $V = I \times R$

$R = \dfrac{V}{I} = \dfrac{12}{0.5} = \mathbf{24\ \Omega}$

[2 marks for correct answer, otherwise 1 mark for correct substitution into the equation.]

iii) The potential difference is split equally across the 12 identical bulbs, so the potential difference across one bulb is $\dfrac{12}{12} = 1$ V *[1 mark]*

b) The total resistance of the circuit was 24 Ω for 12 bulbs. The bulbs are identical, so the resistance of one bulb is 24 ÷ 12 = 2 Ω.
So the total resistance of the new circuit is 10 × 2 = 20 Ω

$V = I \times R$ so $I = \dfrac{V}{R} = \dfrac{12}{20} = \mathbf{0.6\ A}$

[3 marks for correct answer, otherwise 1 mark for calculating the total resistance and 1 mark for substituting the correct values into the rearranged formula.]

2 a) 5 V and 7 V *[1 mark]*

b) Total resistance = $R_1 + R_2 + R_3$
R_3 = total resistance $- R_1 - R_2 = 24 - 2 - 4 = \mathbf{18\ \Omega}$
[2 marks for correct answer, otherwise 1 mark for substituting into the correct equation.]

c) i) $V = I \times R = 0.5 \times 4 = \mathbf{2\ V}$
[2 marks for correct answer, otherwise 1 mark for substituting into the correct equation.]

ii) Because the potential difference across it would not be the same as the potential difference across the component (because potential difference is split across components in series circuits) *[1 mark]*.

Pages 71-72: Parallel Circuits

1 a) M. In parallel circuits, each component is connected separately to the power supply and each can be turned on and off separately *[1 mark]*.

b) i)

Switch 1	Switch 2	Switch 3	Potential difference across heater 1
open	open	open	0 V
closed	open	open	0 V
closed	closed	open	12 V
closed	closed	closed	12 V

[3 marks available — 1 mark for each correct entry.]

ii) The total circuit current splits between the components in parallel, so it is the sum of the currents in the components. Total current = 3 + 3 + 4 = **10 A**
[2 marks for correct answer, otherwise 1 mark for showing understanding that the total circuit current is the sum of the currents through each component.]

iii) The current is the same through series parts of a circuit *[1 mark]*, so it will measure the current through the component if they are in series *[1 mark]*.

c)

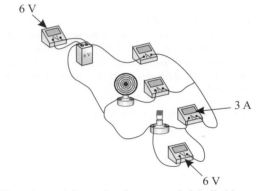

6 V

3 A

6 V

[3 marks — 1 for each value correctly labelled.]

The top voltmeter will show the potential difference of the battery, 6 V. The ammeter will show the current through the heater, which is the current through the fan subtracted from the total current: 4 − 1 = 3 A. The bottom voltmeter will show the voltage across the heater, which is the same as the voltage of the power supply, 6 V.

Page 73: Mains Electricity

1 a) C *[1 mark]*

b) Trace B *[1 mark]*, because the potential difference is constant / it is direct current/d.c. *[1 mark]*.

c) i) Period = time of one full cycle = 4 divisions
= 4 × 0.005 = 0.02 s

$T = \dfrac{1}{f}$ so $f = \dfrac{1}{T} = \dfrac{1}{0.02} = \mathbf{50\ Hz}$

[3 marks for correct answer, otherwise 1 mark for finding the period and 1 mark for correctly substituting it into the correct equation.]

ii) Current that is constantly changing direction *[1 mark]*.

iii) E.g.

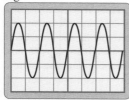

[1 mark for drawing a trace with the correct frequency and with the correct peak voltage.]

iv) 50 Hz *[1 mark]*, 230 V *[1 mark]*

Page 74: Electricity in the Home

1 a) i) X — neutral wire *[1 mark]*
   Y — fuse *[1 mark]*
   Z — live wire *[1 mark]*

   ii) yellow and green *[1 mark]*

   iii) A three-core cable (with a plastic/rubber outer casing) containing three wires (a live, neutral and earth wire), each with a core of copper and a casing of plastic/rubber *[1 mark]*.

   b)

Plug part	Material used	Reason
Plug casing	Rubber or plastic	The plug casing needs to be an insulator.
Plug pins	E.g. Brass/metal	The pins need to conduct electricity well.
E.g. cable grip/cable insulation.	Rubber or plastic	The cables need to be insulated.

[2 marks available — 1 for each row filled in correctly.]

Page 75: Fuses and Earthing

1 a) The vacuum cleaner is double insulated. / Plastic is an (electrical) insulator *[1 mark]*. This means the casing doesn't conduct electricity, so it can never become live *[1 mark]*.

   b) i) The metal casing can become live and give someone an electric shock if they touch it *[1 mark]*.

   ii) A large current surges to earth through the earth wire *[1 mark]*. This causes a large current to surge through the live wire and melt the fuse *[1 mark]*. The fuse breaks the circuit and isolates the microwave oven *[1 mark]*.

   iii) It would need a fuse with a higher current rating *[1 mark]*.

   c) i) When the user touches a live part, current flows through the user towards earth *[1 mark]*. There is a difference between the current in the live and neutral wires *[1 mark]*. The RCCB detects this difference and cuts off the supply *[1 mark]*.

   ii) E.g. RCCBs cut off the current much faster than fuses / RCCBs don't need to be replaced every time they break the circuit *[1 mark]*.

Page 76: Energy and Power in Circuits

1 a) i) So that when the current flows through the wire it gets hot (and so can be used to heat the water in the kettle) *[1 mark]*.

   ii) light *[1 mark]*

   iii) electrical energy to heat energy *[1 mark]*

   b) i) The rate of energy transfer (of an appliance) *[1 mark]*

   ii) Power = 2.5 kW *[1 mark]* and energy transferred in one minute = 180 kJ *[1 mark]*
   Power = energy transferred ÷ time = 150 000 ÷ 60 = 2500 W and energy transferred = power × time = 3000 × 60 = 180 000 J.

   iii) Kettle C *[1 mark — allow mark if wrong value from b) ii) is correctly used.]*

   iv) She should choose kettle A because it has the higher power rating *[1 mark]*. This means that it transfers more energy (to heat energy) per unit time, so it will boil the water faster *[1 mark]*.

Page 77: Power and Energy Change

1 a) $P = I \times V = 2.3 \times 12 = $ **27.6 W**
   [2 marks for correct answer, otherwise 1 mark for correctly substituting into the correct equation.]

   b) No. The fuse should be rated as close as possible but just above the normal operating current *[1 mark]*. If the fuse is below the normal operating current it will blow straight away even if there is no fault *[1 mark]*.

   c) $E = V \times Q = 12 \times 69 = $ **828 J**
   [2 marks for correct answer, otherwise 1 mark for correctly substituting into the correct equation.]

2 a) $P = I \times V$ so $I = \dfrac{P}{V} = \dfrac{184}{230} = $ **0.8 A**
   [2 marks for correct answer, otherwise 1 mark for correctly substituting into the correctly rearranged equation.]

   b) 1 A *[1 mark]*

Page 78: The Cost of Electricity

1 a) On standby: 13598.63 − 13592.42 = 6.21
   Turned off: 13649.41 − 13646.68 = 2.73
   Difference = 6.21 − 2.73 = **3.48 kWh**
   [2 marks for correct answer, otherwise 1 mark for calculating energy used in each case and attempting to subtract.]

   b) i) 34783 − 34259 = 524
   524 × 9.7 = **5082.8p**
   [3 marks for correct answer, otherwise 1 mark for finding energy used and 1 mark for multiplying by 9.7.]

   ii) $E = P \times t = 2.3 \times (180 \div 60 \div 60) = $ **0.115 kWh**
   [2 marks for correct answer, otherwise 1 mark for correctly substituting into correct equation.]

   iii) More energy *[1 mark]*, because it is turned on for longer at the same power *[1 mark]*.

Page 79: The National Grid

1 a)

step-**up** transformer   step-**down** transformer

[2 marks available — 1 mark for each label correct.]

   b) i) Energy is lost as heat due to the heating of the transmission cables *[1 mark]*.

   ii) A step-up transformer is used to increase the potential difference of the electricity supply *[1 mark]*. A high potential difference means a low current, as $P = I \times V$ *[1 mark]*. So there is less heat loss in the power cables and less energy is wasted in transmission *[1 mark]*.

   c) A step-down transformer is used to bring the high voltages down *[1 mark]* to a safe level to use in the home *[1 mark]*.

Section Five — Motors, Generators and Transformers

Page 80: Magnets and Magnetic Fields

1 a) How to grade your answer:
   0 marks: No method is given.
   1-2 marks: A potentially workable method mentioned, but no description of the steps to be taken.
   3-4 marks: Clear description of a workable method including the steps to be taken, but lacking detail.
   5-6 marks: Clear and detailed description of a workable method. The answer has a logical structure and uses correct spelling, grammar and punctuation.
   Here are some points your answer may include:
   Place the four magnets in the position shown on a clear overhead projector/projector slide.

Place many small transparent compasses on the slide in different places between the magnets. / Sprinkle iron filings on the slide between the magnets.
The compasses/iron filings will line up with the magnetic field.
Project the image onto a board or piece of paper.
Use the projected image to trace the shape and position of the four magnets.
Use the projected image to draw the magnetic field by drawing lines between the magnets in the direction that the compasses point. / Use the projected image to draw the magnetic field by drawing lines between the magnets that line up with the direction of the iron filings.

b)   E.g.

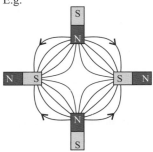

[2 marks for four correct arrows, otherwise 1 mark for two correct arrows, each one placed between a different pair of magnets. Do not award any marks for contradicting arrows.]

c) i)   The field is uniform *[1 mark]*.
   ii)   Attraction *[1 mark]* — opposite poles are facing each other, so there will be a force of attraction between them *[1 mark]*.

Pages 81-82: Electromagnetism

1 a)   a magnetic field is created around the copper rod. *[1 mark]*
   b)

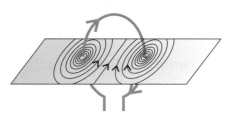

[1 mark for drawing concentric circles centred around the copper rod, 1 mark for drawing arrows on the magnetic field lines pointing anticlockwise.]

c) i)

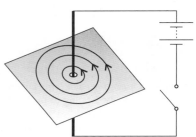

[2 marks available — 1 mark for the pattern and 1 mark for the direction.]
   ii)   The direction of the field would have been reversed *[1 mark]*.
2 a)   A solenoid, or coil of wire (with an iron core) *[1 mark]*.
   b)   When the electromagnet is turned on, current flows through the coil of wire and produces a magnetic field, so the iron is attracted to it *[1 mark]*. When the current stops, there is no longer a magnetic field around the electromagnet *[1 mark]*, so the bar is no longer attracted to the electromagnet and drops *[1 mark]*.
   c)   If the core isn't magnetically soft, it will stay magnetic after the current is turned off, so the crane may not drop the iron bar when it should *[1 mark]*.

3   When the switch is closed, a current flows in the electromagnet and creates a magnetic field *[1 mark]*. The electromagnet attracts the iron arm, causing the striker to hit the bell *[1 mark]*. Meanwhile, the contacts stop touching and the current in the circuit stops *[1 mark]*, so the force of attraction stops and the spring causes the striker to return *[1 mark]*. The contacts touch again, current flows, and the cycle repeats, causing the striker to strike the bell repeatedly *[1 mark]*.

Page 83: The Motor Effect

1 a) i)   A current-carrying wire in a magnetic field experiences a force *[1 mark]*.
   ii)   upwards *[1 mark]*
   b)   The force will increase *[1 mark]*.
   c)   The electric current in the bar is parallel to the magnetic field *[1 mark]*, so the bar will experience no force *[1 mark]*.
2   When the a.c. current flows through the coil of wire in the magnetic field of the permanent magnet, the coil of wire experiences a force *[1 mark]*. The force causes the coil, and so the cone, to move *[1 mark]*. The a.c. current is constantly changing direction, so the force on the coil is constantly changing and so the cone vibrates back and forth *[1 mark]*. The vibrations of the cone cause vibrations in the air which are heard as sound waves *[1 mark]*.

Page 84: The Simple Electric Motor

1 a)   E.g.

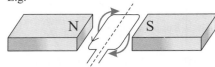

[1 mark for any indication that the current goes anticlockwise.]
   b)   After 90° the force on the top arm will still be acting upwards and the force on the bottom arm will still be acting downwards, so the forces will oppose the rotation of the loop *[1 mark]*.
   c)   By swapping the direction of the current/contacts every half turn (using a split-ring commutator) *[1 mark]*, so the forces on the loop always act in a way that keeps the loop rotating *[1 mark]*.
   d) i)   E.g. increase the current *[1 mark]*. / Increase the strength of the magnetic field *[1 mark]*.
   ii)   Reversing the direction of the magnetic field/swapping the magnets *[1 mark]*. Reversing the direction of the current *[1 mark]*.

Pages 85-86: The Generator Effect

1 a)   As the wheel rotates, the magnet rotates inside the coil of wire *[1 mark]*. This creates a changing magnetic field in the coil of wire which induces a potential difference *[1 mark]*.
   b)   Any three of: By increasing the strength of the magnet *[1 mark]*. / By increasing the number of turns on the coil of wire *[1 mark]*. / By increasing the speed of rotation of the magnet (e.g. by using a faster hamster / a smaller hamster wheel) *[1 mark]*. / By increasing the area of the coil of wire *[1 mark]*.
   c)   When the magnet is turned through half a turn, the direction of the magnetic field reverses *[1 mark]*, so the direction of the induced potential difference reverses and the current flows the opposite way round the coil *[1 mark]*.
   d)   there will be a potential difference induced across the ends of the wire. *[1 mark]*
2 a)   The vibrations of the sound waves cause the diaphragm, and so the coil of wire, to move forwards and backwards *[1 mark]*. The coil of wire is moving in relation to the fixed magnet, so the coil of wire cuts the magnetic field *[1 mark]*. This causes a potential difference to be induced across the coil *[1 mark]*.

b) i) Faster vibrations would cause the coil to change direction more frequently *[1 mark]*, so the induced potential difference would change direction more frequently *[1 mark]*.
  ii) The peak potential difference would increase *[1 mark]*.

Page 87: Generators

1 a) Rotating the handle causes the coil to rotate (move) within the magnetic field, so a potential difference (and so current) is induced in the circuit *[1 mark]*. The direction of the wire's movement in relation to the magnetic field changes every half turn and so the direction of the current/potential difference induced changes *[1 mark]*.

b) E.g.

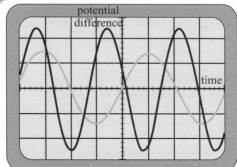

[2 marks available — 1 mark for drawing a trace with a larger difference between maximum and minimum potential difference than the original trace and 1 mark for drawing a trace with a higher frequency than the original trace.]

c) The trace would be the same as when the coil is rotating and the magnets are fixed *[1 mark]*.

d) The slip rings should be replaced with something (a split-ring commutator) that causes the electrical contacts to swap every half turn *[1 mark]*, so that the current continues to flow in the same direction *[1 mark]*.

Pages 88-89: Transformers

1 a) A transformer is made up of two coils of wire, the primary coil and the secondary coil *[1 mark]*, wound round a soft iron core *[1 mark]*.

b) a changing magnetic field *[1 mark]*

c) Step-down, because the potential difference decreased (from 240 V to 12 V) *[1 mark]*.

d) If the transformer is 100% efficient, $V_p \times I_p = V_s \times I_s$.

$$I_s = \frac{V_p \times I_p}{V_s} = \frac{240 \times 0.25}{12} = 5\,\text{A}$$

[2 marks for correct answer, otherwise 1 mark for substituting correctly into rearranged equation.]

e) E.g. the laptop transformer operates at a much higher frequency than the spotlight transformer (around 50-200 kHz) *[1 mark]*. The laptop transformer is much lighter and smaller than the spotlight transformer *[1 mark]*.

2 a) The alternating current flowing in the primary coil produces a changing magnetic field in the iron core *[1 mark]*. The changing magnetic field in the iron core cuts through the secondary coil *[1 mark]*. A potential difference is induced in the secondary coil by the changing magnetic field *[1 mark]*, which causes an alternating current to flow *[1 mark]*.

b) i) $\frac{V_p}{V_s} = \frac{n_p}{n_s}$ so $V_s = \frac{n_s}{n_p} \times V_p = 16 \times 25\,000 = \mathbf{400\,000\,V}$

  [3 marks for correct answer, otherwise 1 mark for using $\frac{n_s}{n_p} = 16$, and 1 mark for substituting correctly into the rearranged equation.]

  ii) The transformer is 100% efficient, so input power = output power.
  So $V_p \times I_p = 4\,000\,000$ and the input voltage is $V_p = 25\,000\,V$
  So $I_p = 4\,000\,000 \div 25\,000 = \mathbf{160\,A}$
  [2 marks available for correct answer, otherwise 1 mark for equating the output power with $V_p \times I_p$.]

Section Six — Nuclear Physics

Page 90: Atomic Structure

1 a) i)

Particle	Relative charge	Relative mass	Number present in an atom of iodine-131
Proton	1	1	53
Neutron	0	1	78
Electron	−1	very small	53

[4 marks — 1 for each correct answer.]
  ii) protons and neutrons *[1 mark]*
  iii) electrons *[1 mark]*

b) Atoms with the same atomic number but a different mass number *[1 mark]*.

c) i) 0 (no overall charge) *[1 mark]*.
  ii) ion *[1 mark]*
  iii) No effect *[1 mark]*, because the mass number and atomic number only tell you about the numbers of protons and neutrons, which don't change *[1 mark]*.

Page 91: Radiation

1 a) at random. *[1 mark]*
b) i) background radiation *[1 mark]*
  ii) Any two from: e.g. nuclear weapons tests *[1 mark]* / cosmic rays from space *[1 mark]* / nuclear accidents *[1 mark]* / nuclear waste *[1 mark]*.

2 a) E.g. alpha (particles), beta (particles) and gamma (rays) *[1 mark]*.
b) How to grade your answer:
  0 marks: There is no relevant information.
  1-2 marks: There is a brief description of the danger posed by (some types of) radiation.
  3-4 marks: There is some description of the dangers posed by radiation, with reference to the different levels of danger posed by different types of radiation. The answer has a logical structure, and spelling, punctuation and grammar are mostly correct.
  5-6 marks: There is a clear and detailed description of how radiation can be harmful, and a clear description of the different situations in which each type of radiation is more or less dangerous. The answer has a logical structure and uses correct spelling, grammar and punctuation.
  Here are some points your answer may include:
  Each type of radiation causes ionisation of living cells, which can damage or destroy them.
  Minor cell damage caused by radiation can give rise to mutant cells, which can lead to cancer.
  High doses of radiation can kill body cells completely and cause radiation sickness.
  Outside the body beta and gamma sources are the most dangerous because they have the most penetrating power.
  Outside the body, alpha sources are not very dangerous because they can't penetrate the skin very much.
  Inside the body, alpha sources are the most dangerous because they do all their damage in a very localised area.
  Inside the body, beta and gamma sources are not as dangerous because many of them pass straight out and they are less ionising than alpha sources.

Pages 92-93: Ionising Radiation

1 a) Alpha (particles) *[1 mark]*, because they are relatively large and heavy, so they collide with lots of atoms, causing ionisation *[1 mark]*.
b) i) gamma (rays) *[1 mark]*
  ii) beta (particles) *[1 mark]*

2 a) An alpha particle is made up of 2 protons *[1 mark]* and 2 neutrons (it's a helium nucleus) *[1 mark]*.

b) The atomic number decreases by 2 *[1 mark]* and the mass number decreases by 4 *[1 mark]*.

c) $^{206}_{84}\text{Po} \rightarrow ^{202}_{82}\text{Pb} + ^{4}_{2}\alpha$

[2 marks — 1 for each correct number.]

Remember, the mass number, at the top left, gives the number of protons and neutrons and the atomic number, at the bottom left, gives the number of protons only. So when a nucleus releases an alpha particle, its mass number reduces by 4 and its atomic number goes down by 2.

3   alpha and gamma *[1 mark]*
    Alpha particles are completely stopped by both paper and aluminium, which explains the count rate dropping to half *[1 mark]*. Beta particles are stopped by aluminium, so the other type of radiation must be gamma (which passes straight through both) *[1 mark]*.

4 a) i) A gamma (radiation) *[1 mark]*
            B alpha (particles) *[1 mark]*
            C beta (particles) *[1 mark]*
       ii) A / gamma *[1 mark]*
    b) electric field *[1 mark]*
    c) E.g. The two types of radiation move in opposite directions *[1 mark]*. This is because they have opposite charges *[1 mark]*. The alpha radiation (B) is deflected less than the beta radiation (C) *[1 mark]*. This is because alpha particles have a greater mass than beta particles *[1 mark]*.

Pages 94-95: Half-Life

1 a) To get a measure of the background radiation *[1 mark]* so that the processed results can take it into account *[1 mark]*.
    b) 30 minutes *[2 marks for correct answer, otherwise 1 mark for evidence of using the graph correctly to find the half-life]*.

To find the half-life, just look at the point on the curve where the count rate is half its original value, i.e. 400 counts per second.

    c) Number of half-lives elapsed after 120 minutes = $\frac{120}{30}$ = 4 *[1 mark]*
        After 4 half-lives, count rate will be 800 ÷ 2 ÷ 2 ÷ 2 ÷ 2
        = **50 counts per second** *[1 mark]*.
    d) The older sample would have a lower count rate to begin with *[1 mark]*. This is because the half life will remain the same *[1 mark]*, but more of its nuclei will have already decayed so it will emit less radiation *[1 mark]*.

2   decrease *[1 mark]*, nuclei *[1 mark]*, halve *[1 mark]*.

3 a) i) 2 × 60 = 120 minutes
            120 ÷ 40 = 3 half-lives
            8000 ÷ 2 = 4000, 4000 ÷ 2 = 2000, 2000 ÷ 2 = **1000 cps**
            [2 marks for the correct answer, otherwise 1 mark for calculating the number of half-lives in 2 hours.]
       ii) 8000 ÷ 2 = 4000, 4000 ÷ 2 = 2000, 2000 ÷ 2 = 1000,
            1000 ÷ 2 = 500, 500 ÷ 2 = 250, 250 ÷ 2 = 125. So it takes 6 half-lives to drop to less than 200 cps.
            6 × 40 = 240 mins
            240 ÷ 60 = **4 hours**
            [2 marks for the correct answer, otherwise 1 mark for calculating the number of half-lives taken to reach a count rate below 200 cps.]
    b) B *[1 mark]*

Pages 96-97: Uses of Radiation

1 a) The gamma rays kill any potentially harmful microbes *[1 mark]*.
    b) i) Technetium-99m *[1 mark]*
            This isotope emits gamma rays, which mostly pass straight through the body without damaging cells *[1 mark]*, and its half-life is the right length to be able to carry out tests without the patient being radioactive for a long time after *[1 mark]*.
       ii) High doses of ionising radiation kill living cells *[1 mark]*, so the radiation can be directed towards cancer cells to kill them *[1 mark]*.

iii) Ionising radiation from radioactive isotopes also kills or damages healthy cells *[1 mark]*, which can cause serious health problems *[1 mark]*.

2 a) Alarms work by using a radioactive source to ionise air inside the alarm *[1 mark]*. Alpha radiation would be absorbed by the smoke in the event of a fire, beta and gamma radiation wouldn't and so the alarm would not work *[1 mark]*. / Alpha radiation is strongly ionising, so would ionise the air, allowing the alarm to work *[1 mark]*. Beta and gamma radiation are less strongly ionising, so may not ionise the air enough for the alarm to function properly *[1 mark]*.
    b) The range of alpha particles in air is quite short, so a large distance would mean not many alpha particles reach the air between the electrodes *[1 mark]*.
    c) Americium-241 *[1 mark]*. It emits alpha radiation *[1 mark]*, and has a very long half-life, so wouldn't need replacing very often/ever *[1 mark]*.

Page 98: Nuclear Fission and Fusion

1 a) Uranium-235 or plutonium-239 *[1 mark]*
    b) i) A slow-moving neutron gets absorbed by an (unstable) nucleus *[1 mark]*. The nucleus splits to form two daughter nuclei *[1 mark]*, releasing two or three neutrons *[1 mark]* and a large amount of energy *[1 mark]*.
       ii) E.g.

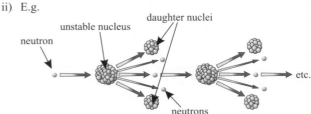

[3 marks available — 1 mark for showing a neutron colliding with an unstable nucleus, 1 mark for showing two daughter nuclei and two or three neutrons and 1 mark for showing at least one more identical fission reaction.]

2 a) Nuclear fusion *[1 mark]*
    b) Hydrogen nuclei fused together in stars to form larger nuclei *[1 mark]*, which in turn fused together to form even larger nuclei *[1 mark]*. Elements heavier than iron are formed when stars explode (supernovas), which also cause the distribution of all of the elements throughout the Universe *[1 mark]*.

Page 99: The Life Cycle of Stars

1 a) i) gravitational force / gravity *[1 mark]*
       ii) E.g. planets *[1 mark]*
    b) i) E.g. The forces in the star are balanced *[1 mark]*. / The outwards pressure/force (caused by nuclear fusion/heat) balances the inwards (gravitational) force *[1 mark]*.
       ii) Stars can output energy for millions of years because they contain vast amounts of hydrogen that can be fused to release energy *[1 mark]*.
    c) i) The star swells into a red giant *[1 mark]*, which (ejects its outer layers and) becomes a white dwarf *[1 mark]*, which eventually cools to a black dwarf *[1 mark]*.
       ii) A: Red super giant *[1 mark]*
            B: Neutron star *[1 mark]*
            C: Black hole *[1 mark]*
       iii) supernova *[1 mark]*

Equations Pages

Here are some equations you might find useful when you're doing the practice papers — you'll be given these equations in the real exams.

Section One — Forces and Their Effects

v = velocity, s = displacement, t = time	$v = \dfrac{s}{t}$
a = acceleration, v = final velocity, u = initial velocity, t = time taken	$a = \dfrac{v - u}{t}$
F = force, m = mass, a = acceleration	$F = m \times a$
p = momentum, m = mass, v = velocity	$p = m \times v$
F = force, Δp = change in momentum, t = time	$F = \dfrac{\Delta p}{t}$
W = weight, m = mass, g = gravitational field strength (acceleration of free fall)	$W = m \times g$
W = work done, F = force, d = distance moved in the direction of the force	$W = F \times d$
E_p = change in gravitational potential energy, m = mass, g = gravitational field strength (acceleration of free fall), h = change in height	$E_p = m \times g \times h$
E_k = kinetic energy, m = mass, v = speed	$E_k = \frac{1}{2} \times m \times v^2$
F = force, k = spring constant, e = extension	$F = k \times e$
P = power, W = work done, t = time	$P = \dfrac{W}{t}$
M = moment of the force, F = force, d = perpendicular distance from the line of action of the force to the pivot	$M = F \times d$
T = time period, f = frequency	$T = \dfrac{1}{f}$
P = pressure, F = force, A = cross-sectional area	$P = \dfrac{F}{A}$

Section Two — Waves

v = speed, f = frequency, λ = wavelength	$v = f \times \lambda$
n = refractive index, i = angle of incidence, r = angle of refraction	$n = \dfrac{\sin i}{\sin r}$
n = refractive index, c = critical angle	$n = \dfrac{1}{\sin c}$
s = distance, v = speed, t = time	$s = v \times t$

magnification = $\dfrac{\text{image height}}{\text{object height}}$	
P = power of a lens, f = focal length	$P = \dfrac{1}{f}$
u = object distance, v = image distance, f = focal length	$\dfrac{1}{u} + \dfrac{1}{v} = \dfrac{1}{f}$

Section Three — Heating Processes

E = energy, m = mass, c = specific heat capacity, θ = temperature change	$E = m \times c \times \theta$
E = energy, m = mass, L_v = specific latent heat of vaporisation	$E = m \times L_v$
E = energy, m = mass, L_F = specific latent heat of fusion	$E = m \times L_F$
efficiency = $\dfrac{\text{useful energy out}}{\text{total energy in}}$ (× 100%) OR $\dfrac{\text{useful power out}}{\text{total power in}}$ (× 100%)	

Section Four — Electricity

I = current, Q = charge, t = time	$I = \dfrac{Q}{t}$
V = potential difference, E = energy transferred, Q = charge	$V = \dfrac{E}{Q}$
V = potential difference, I = current, R = resistance	$V = I \times R$
T = time period, f = frequency	$T = \dfrac{1}{f}$
P = power, E = energy transferred, t = time	$P = \dfrac{E}{t}$
P = power, I = current, V = potential difference	$P = I \times V$
E = energy transferred, V = potential difference, Q = charge	$E = V \times Q$
E = energy transferred, P = power, t = time	$E = P \times t$

Section Five — Motors, Generators and Transformers

V_p = potential difference across the primary coil, V_s = potential difference across the secondary coil, n_p = number of turns on the primary coil, n_s = number of turns on the secondary coil	$\dfrac{V_p}{V_s} = \dfrac{n_p}{n_s}$
V_p = potential difference across the primary coil, I_p = current in the primary coil, V_s = potential difference across the secondary coil, I_s = current in the secondary coil	$V_p \times I_p = V_s \times I_s$